The Dance of the Arabian Babbler

The Dance of the Arabian Babbler

Birth of an Ethological Theory

Vinciane Despret

Translated by jeffrey bussolini

A UNIVOCAL BOOK

University of Minnesota Press
Minneapolis | London

Published by the University of Minnesota Press
111 Third Avenue South, Suite 290
Minneapolis, MN 55401-2520
http://www.upress.umn.edu

ISBN 978-1-5179-1152-2 (pb)
Library of Congress record available at https://lccn.loc.gov/2020058456.

Printed in the United States of America on acid-free paper

The University of Minnesota is an equal-opportunity educator and employer.

30 29 28 27 26 25 24 23 22 21   10 9 8 7 6 5 4 3 2 1

To Amotz Zahavi
for accompanying and guiding me in
my discoveries

To Isabelle Stengers
for guiding and then accompanying me in
my rediscoveries

Contents

Acknowledgments

The trip and the stay in Israel that are at the origins of this text were able to be carried out thanks to the generosity of many sources of assistance and many people. I would therefore like to express my gratitude to the French Community of Belgium, which offered me the "Travel Award," and the FNRS and the endowment of the Université de Liège, which gave me a subsidy allowing me to prolong my stay. Without them, this project would simply have remained a dream. Those who should equally be thanked for the confidence they had the goodwill to accord me are Danielle Bajomée, president of the Department of Philosophy and Communication Sciences, and André Motte, dean of the Faculty of Philosophy and Letters, both at the Université de Liège. They not only encouraged my project, they also gave me all the extended leave I needed to be able to realize it.

In Israel, the kindness and willingness of Amotz Zahavi, Jonathan Wright, and Roni Osztreiher were a tremendous boon to my work.

My family, Jean-Marie and Jules-Vincent Lemaire, and Jacques and Michou Dubois gave me significant help, as much at the level of organization of the trip as in offering me their benevolent care and their assistance.

I would also like to thank for their encouragement and their critical advice Abdessadek El Ahmadi, Ninette Grooteclaes, Jean-Marie Lemaire, Francis Pire, Jean-Claude Ruwet, my editor, Phillipe Pignarre, for his confidence and the work of rereading, and above all Isabelle Stengers, without whom this book would have been able to take neither its breath nor its flight.

Translator's Introduction

jeffrey bussolini

In her first monograph Vinciane Despret charts a fascinating course that interfaces with some of the most decisive arguments in ethology and the social studies of science while also bringing creativity, insight, and complexity to the issue at hand. That issue, of course, is the peculiar and wonderful birds called Arabian babblers who were the objects of study of the equally peculiar and wonderful biologist Amotz Zahavi, who devoted his professional life—and indeed a part of his personal and social life—to their interpretation, companionship, and conservation. The particulars of the story are important. Although Despret was familiar with Zahavi's published work and the concepts and debates it engaged, she did not carry out an armchair study, but realized that, like Zahavi, she would have to go "into the field" herself to observe and take part in the contexts of interaction. Following previous ornithological ethology that she had carried out on the Western capercaillie with Jean-Claude Ruwet, Despret received a research travel grant that allowed her to go to Israel and observe Zahavi (and others) observing the babblers—although as she notes this research on both sides was not a distanced or "outsider" endeavor, but a study from within the social milieu of the birds and the researchers. Indeed, she credits going to the field and taking part in that kind of immanent (participant) observation with dramatically reframing the questions of research she had originally brought to the topic.

The encounter with Zahavi and the babblers was fortuitous,

and changed the course of the research, but it was not accidental. Despret had noted that Zahavi was making unique interventions in the literature on evolutionary biology that were producing effects at the heart of the discipline: as she details in this book, Richard Dawkins had ridiculed Zahavi's ideas in the first edition of his widely influential *The Selfish Gene,* only to have to backtrack and indicate the plausibility and importance of Zahavi's work on signaling theory and the handicap principle in the second edition. She surmised that something about the peculiarity of these birds and this researcher was indispensable to the types of theories that Zahavi was able to create. This also set up one of her guiding questions for the research: was the uniqueness demonstrated in their behavior, or in the distinctive gaze with which Zahavi was viewing them?

The answer, perhaps unsurprisingly, is both: the babblers and Zahavi constitute a central relational "hybrid," drawing on Bruno Latour's concept that he describes as a mediation between nature and culture, where the actors cue and modify one another. But it was not foreordained that Despret would recognize the uniqueness and activity of each party to the hybrid. As she says, she had gone to the field with the notion of performing a fairly direct transcription of human social conditions to the results of the scientific work. She conjectured, for instance, that the particular geopolitical and economic conditions of Israeli society might directly account for the reading of the babblers. What she ultimately found, while not forsaking some of those insights, was much more complex and much more interesting.

But Despret's emphasis on the hybrid is not to say, either, that the originality of the babblers is anything less than astonishing. For one thing, they have a distinctive practice that attracted both Zahavi and Despret to studying them: they dance. The act of dancing leads to questions that have wide reverberations through biology and the social sciences: If the birds expend energy and make themselves vulnerable to predation, why do they dance? And, in calling this activity "dancing" and considering it in relation to other ritual behavior among non-human animals, how are cultural, as well as evolutionary, continuities between nonhuman and human animals framed? So, the babblers' dancing, gift giving, helping other birds at the nest, general social malleability, and complex social structures and interactions become key behaviors for Zahavi and the other humans fascinated by these birds. One of the most decisive aspects of Zahavi's approach, according to Despret, was the way that he left so much latitude to the birds in determining the important questions and key behaviors to be focused on. Put simply, his task is merely to learn to follow them, to be able to decipher the complexity of some of their behavior.

Despret credits Zahavi's "anthropological approach," which sees the birds as variable and interpreting actors rather than passive reactors to environmental stimuli, with opening up the possibility of his being able to interpret the importance and beauty of behaviors like dancing. She calls this anthropological approach of Zahavi's an a posteriori one that does not come to the field with specific hypotheses to test, but lets the animals guide and frame the questions to be addressed. She contrasts this a posteriori approach with an a priori one that establishes particular hypotheses to test before coming to the field, and that modifies aspects of animals' appearance or environment to look for variations from a norm. Behind this model is an ethological tradition that sees animal behavior as preprogrammed responses to particular environmental stimuli. Part of Zahavi's discontent with that approach was, simply, that he saw the babblers react in a variety of ways to the "same" stimuli: as Despret puts it, he started to discern difference where similarity had been assumed before. Despret, too, swapped an a priori approach based on a hypothesis of direct social construction for a messier and more multifactorial a posteriori approach when confronted with the complexity of the actions and interactions in the field. In characterizing the stakes of the differences between the a priori and a posteriori approaches, Despret describes them as the difference between a trial, in which the parties stand accused of particular charges or expectations that are stated in advance, and an investigation, which is a more open-ended attentiveness to the situation and the course of events without a preordained presumption.

A central aspect of Zahavi's a posteriori, anthropological approach is a practice that also lies at the heart of Despret's: telling stories. She believes that his method of charting the individual birds' actions and activities over the span of many years to track their social interactions, the "gaze" he directs toward the babblers, gives him entrée to some aspects of the babblers' world and the insight to interpret their peculiar behavior. Telling stories initially perplexed some of Zahavi's colleagues and competitors—Dawkins wondered whether data related in words alone could match up to the rigor of mathematical models as in the game theory of evolution—but ultimately underscored the place for anecdotal accounts in scientific research (as Marc Bekoff and many others have emphasized). Likewise, Despret focuses on both Zahavi's approach and on the telling of stories as a general, philosophical frame for studying the process of research and thinking—in the natural and social sciences as well as the humanities. In this respect she is not only addressing pivotal debates within biology, but also immersed in the cutting-edge developments and approaches in the philosophy and sociology of science. It is this approach, too, that I believe drew Despret's interest to our work on feline sociality at the Center for Feline Studies and prompted her to ask me to translate this book.

As she indicates, Despret draws on concepts and frames of reference from Isabelle Stengers and Bruno Latour. Although she relies heavily on their writings and perspectives, in this book she especially expands on Stengers's ideas about the philosophical disposition behind scientific approaches (including the importance of laughter) and the ways that experiments produce existence rather than just revealing it, and Latour's notions of the hybrid, as noted previously, and fictionalization in scientific practice. As in the work of Stengers and Latour themselves, both noted scholars of the practice and development of science, "production" and "fiction" here do not mean that the results of science are relative and entirely detached from physical and biological reality: indeed, it is the fact of the babblers' dancing that serves as the point of departure and return for the inquiries here. Rather, these concepts pay heed to how particular scientists and contexts are involved in the unfolding of scientific research and theories, and how the practices and results of scientific work are narrativized to fit within or contest dominant paradigms. Beyond a naïve realism or a relativistic constructivism, the approach here is better characterized by what philosopher of science Babette Babich has called "perspectivalism" in her study of *Nietzsche's Philosophy of Science,* where grounding, disposition, and activity matter. It should also be noted that the technique and the movement of Despret's book show why she and Donna Haraway have had such an ongoing and productive exchange of ideas and friendship. Despret's combination of fieldwork, scientific literacy, critical practice, and philosophical attention to the implications of method make them kindred spirits and scholars.

The major application of fictionalization here is to look at how experiments and the process of scientific publication serve as contexts for the adjudication of competing fictions. Different methods, different field approaches, and different ways of playing, or opting out of, the "scientific game" play roles in the acceptance, rejection, or ignoring of research. Zahavi and the babblers are perfect interlocutors to consider this dynamic: the birds offer perplexing behaviors that seemed to contradict evolutionary reasoning and open onto culture, so Zahavi adopted an anthropological, storytelling approach that shifted the way their activities and interactions would be narrativized. No wonder the babblers and the eccentric ornithologist provoked such indigestion in the scientific community, as they pointed out dimensions that needed to be rethought and reframed in evolutionary theory. Adding onto this, Despret points out how Zahavi's "refusal to play the scientific game" as conventionally determined affected the reaction to his theory.

Location and grounding matter. It was not an accident that these behaviors and ideas emanated from outside of the European context. Indeed, Despret writes that Zahavi's position outside the European mainstream at the University of Tel Aviv, studying birds from the Negev desert, gave him a latitude for observation and theory development that he would not have had at a center of ethological knowledge like the University of Oxford (and helps to explain the initial resistance to his theories from those gatekeepers). But the considerations about location and landscape extend farther than that, and double back on the considerations about the influence of social and geographic context on scientific work. As part of this dimension, she recalls that, though his data came from many sites visited during his voyage on the *Beagle*, Darwin also draws many examples from his native England, and that his writing and thinking was produced within the sociopolitical— and economic—context of nineteenth-century conditions in that country. While he warned against the excessive interpretations of his work as in the social Darwinism of Spencer, Despret points to the ascendant economic liberalism of the time as resonant with the idea of competition and the struggle for survival. As a counterpoint to this she figures the work of Pierre Alexander Kropotkin, the naturalist and anarchist who emphasized cooperation and mutual aid rather than competition. However, in keeping with the grounded, ecophysiological account that she prefers over simple social constructivism, it was not merely the case that Kropotkin saw what he wished to bolster utopian beliefs, but that the *difference* of the land and the ecosystems in Siberia also gave him access to different information, made him part of a different set of hybrids.

The conceptual lines between Darwin, Kropotkin, naturalism, and economics point to deep-seated crossovers between economics and evolutionary biology that this book addresses. Indicating the enduring status of this tie, *Freakonomics* author

Stephen Levitt has recently said, reminiscing on his under-graduate study with E. O. Wilson, that there are no two disciplines with a closer tie than evolutionary biology and economics. Despret highlights the idea in biology, proposed, for instance, by Konrad Lorenz in what she calls his "Lorenzian anthropology," that competition not only guarantees survival but serves as the basis for social ties and pro-social behavior in animal groups. Consonant with this, the concept of cost emerges and takes on a major role, so that animal behaviors and sacrifices come to be reckoned in terms of a quasi-monetary transaction that belies the influence of something more than the behaviors observed themselves—namely, human social ideologies. This seems to show more about the borrowing of ideas between these fields of study than it does about the natural status of these traits and behaviors, as biologist Richard Lewontin expressed in his concept of "secondary derivation." And as both Despret and Lewontin point out, the ideas clearly travel back and forth, since economic ideas used to interpret animals can in turn use that animal behavior itself as proof for the natural sanctity of those economic ideas.

As we have adumbrated here, Despret tells the story of her research in relation to the unfolding of the biologists' research in the field. Methodologically, as we have pointed out, this concretizes the reflexivity "all the way down" that she has identified as being at the perspectival, philosophical heart of both undertakings. Just as she points to the importance of Zahavi's grounding with the babblers and in Tel Aviv, so also grounding matters to her thought and work. For one thing she refers to her situated knowledge within a francophone intellectual tradition as allowing for insights and concept generation that would not have been as easily accessed in an anglophone setting with its attendant traditions and constraints (we might think here of her serious use of the concepts of Deleuze and Guattari in this book at a time when much of the citation of their work in anglophone contexts tended toward less careful and more flippant appropriations, as well as of the aforementioned ongoing dialogue with Stengers and Latour). Within that francophone setting, she is located in Belgium—curiously both peripheral and indispensable to French intellectual life—and has spent most of her career at the Université de Liège, accessing and extending the particular currents of philosophical and psychological research there, as well as undertaking the ethological experience mentioned earlier with Jean-Claude Ruwet. But, given the influence of her research and thinking, Despret has also long taught at Stengers's institution, the Université libre de Bruxelles, in ethology and sociology (which should be little surprise given the topics covered in this book). Noting her intellectual territories and trajectories gives us a better sense of the elements and particularities of her work and hybrids, just as she does for Darwin, Kropotkin, and Zahavi in terms of their relations to particular territories and specific intellectual traditions.

I would like to thank Joan Stambaugh, who originally taught me the craft and stakes of translation, and Vinciane Despret, who asked me to translate this book and has been an inspiring interlocutor about matters philosophical, ethological, feline, and culinary. I thank Donna Haraway, Erica Fudge, Isabelle Stengers, Cary Wolfe, and Bruno Latour, who encouraged me to pursue the translation of Vinciane's work (and advocated for the translation of her work), and Dominique Lestel, who introduced me to her in the first place. Babette Babich, Swapna Mukherjea, Ananya Mukherjea, Jacques Derrida, and Michel Tibon-Cornillot have long encouraged me to attend to both sides of the Great Divide between nature and culture—or indeed to see it as Vinciane does, as a field of continuities rather than a divide at all. This translation would not have been possible without Ananya's meticulous reading, reflection, and assistance. Mark Roth and Andrés Marino facilitated the translation and discussed important concepts with me. I also would like to thank Brett Buchanan and Matthew Chrulew, with whom I have been undertaking a long-term series of projects on philosophical ethology, and Drew Burk, who read the translation and offered invaluable suggestions.

Introduction

The Arabian Babbler (*Turdoides [Argya] squamiceps*) is a
member of the Paleotropic family Timallidae [Leiothrici-
dae]. . . .[1] [Only a few forms of the genus Turdoides [Argya]
have penetrated into the Palaearctic arid zone north of the
tropics. . . .] These Palaearctic forms are distributed over hot
deserts from India to Morocco and south to the arid deserts
of East Africa. *T. [Argya] squamiceps* occurs in the Arabian
and Sinai peninsulas [extending into the hot deserts of
Israel.] In Israel the bird is common along the Rift Valley,
north to Jericho. . . . The Arabian Babbler weighs 65–85 g;
it is about 280 mm long, of which over half (145–55 mm) is
tail. Its color is a dull gray-brown and very cryptic against
the desert background. Babblers are quite terrestrial and
hop and walk more than they fly. [When they fly, they do so
slowly and consequently are vulnerable to predation when
they cross open terrain. This may be the reason why they
usually stay near bushes and trees, where they may escape
into cover, with their strong legs and long tail enabling
them to dodge and outmaneuver any predator. They usually
fly near the ground, low-flying birds being more difficult to
locate and also a more difficult target for a swift potential
predator.] The diet of the Arabian Babbler consists mainly of
small animals, mostly arthropods, found on the ground or
within the vegetation.[2]

The article by the Israeli ethologist Amotz Zahavi starts
more or less in the same manner as the articles by his American and English colleagues working and publishing in this

area. The birds themselves, at least in view of these first few lines, do not seem to bring anything particularly original to ornithological diversity. Only their English name could potentially lead one to crack a smile: the *babbler*, a pretty name meaning chatterbox.

But after these few conformist lines describing a bird that seems to be much the same as any other, the article takes a shocking turn: in the place of figures expressing the overall work of the group—how many young were raised, how many survived, how many eggs, how many birds help their parents around the nest—a first table traces the life of each of the individuals in a group of fifty birds. For each of these babblers are recorded, in the form of different symbols and for each year between 1972 and 1987, the different events that mark their fifteen years of life: their departure from the parental territory, their first eggs, their companions, those who help them in raising young, if they themselves sometimes assume that role, their search for territories, their failures and successes, et cetera. The rupture with the traditional style does not stop there: in the place of statistics, residues of experiments proving hypotheses, in the place of tables with pure lines, the entire article tells stories, as it were: here is what the birds do in the course of their life, and here is no doubt why they do it that way. To this particular "saying" corresponds a "doing" that is at least as surprising: quite evidently, the babbler bird was not a bird like the others.

For the relatively monotonous behaviors usually described in this genre of literature are substituted, in Zahavi's article, the most astonishing descriptions. The babblers live in groups of three to fifteen individuals, of which only one pair, *in principle,* reproduces. This is, among birds, really nothing exceptional. But the term "in principle" should draw our attention. This privilege of the reproductive couple is no more an invariant of the program than a previously inscribed rule. It represents, rather, a compromise that certain groups seem to reach by actively creating it, as if it were necessary to reinvent it each time. And like all active creation, all invention, these compromises can take diverse and original forms, give a particular tonality to the relations between the dominants and the dominated, and mobilize a fascinating behavioral range among the birds: the babblers play, bathe together, offer one another presents, offer to do favors for one another, preen one another, can feed their congeners, and even seem to sometimes enter into conflict for the privilege of helping or giving to others. Finally, they dance in a group at certain times of the day, sometimes for as long as thirty minutes. From these amazing "stories" followed theories that were no less so. We will come back to them.

After reading this article, my questions had to change.

At the time, two questions possessed me. The first concerned our tendency to be moral: could this tendency be partially due to a phylogenetic heritage? I subsequently searched the ethological literature for anything that seemed to me able to constitute the precursors for our moral behaviors and more particularly for our altruistic behaviors.[3] To this first question another corollary question was added: can we, on the basis of these analogies, establish a line of continuity between human and animal?

Birds, as we will have the opportunity to explain at greater length, show many behaviors that are identified as altruistic, most often in playing, for congeners, the role of helper at the nest. They bring food to the brood, can sometimes (even if rarely) incubate the eggs, and above all provide a real contribution to the defense of territory. The activities identified as altruistic are generally reduced to one or another of these behaviors, and the articles are limited to testing one or another explanatory hypothesis regarding their subject. If we want to know a greater diversity of social behaviors, it is necessary to turn toward the literature dedicated to mammals, or better still—as a subset among them—to primates.

Reading Zahavi's article rerouted me: the birds that he described behaved like mammals. Were they therefore so special in having given rise to both surprising observations and inventive and marginal theories? Or was it, rather, necessary to adopt the inverse hypothesis: that it was not so much the birds who were extraordinary, but the gaze that Zahavi addressed to them that conferred the exceptional qualities? Were the babblers flying monkeys—as I had then called them—or was it rather Zahavi who had exaggerated his observations with the craziest interpretations and transformed elementary behaviors into complex games?

Then the question of knowing whether the animal "is moral" was replaced with the question of knowing why some authors define it as moral—and observe many behaviors analogous to moral behaviors—where other authors—with relevant supporting observations—describe them as egoistic, even bloodthirsty.

For the question of continuity or rupture—the question of the meaning of the precursor—is substituted the question of the gaze that creates this continuity or rupture. We know well that none of our representations can be considered neutral. The representations that we build when we have animals in mind are generally even less so. Being like an animal can take on, according to the culture—and according to the animal considered—the most positive or the most negative valences, can represent the most desirable or the most worrying of experiences. The animal became, in Western cultures and particularly in the Enlightenment, the place of "projections" in the psychoanalytic sense of the term:[4] the nature that frightens from the outside, once mastered, becomes a menace from the inside.

Animals offer humans a more or less deforming, more or less satisfactory mirror: both near and far, both same and different. The gaze that we bring to bear on them, the way in which we construct the representation, can render them more "same" or more "different." Similarly, animals are at once object and subject. And as object–subject, they *define* at the same time as they *constitute* the human as subject–subject: they define it in identity and constitute it in otherness. It is there that the fictional role of the precursor plays out: being at the same time same and different.

My question from then on became: What is there, then, to this discourse that we cling to on the subject of this sameness–difference? How does this discourse participate in this identity and this difference? In other words, when we see an animal, what can we say about our "seeing"? If what Jean says about Pierre frequently teaches us much more about Jean than Pierre,[5] what does our ethological discourse teach us about ourselves? Does what Zahavi tells us about the babblers teach us more about Zahavi or about his babblers? Trying to understand what, in Zahavi's ethological discourses, pertains to Zahavi's gaze or to the singularity of the babbler leads us to the question of continuity or rupture: how do we arrive at thinking that an animal is the "same," how do we arrive at thinking that it is "different"? Here I refer explicitly to the question of "how" and not the question of "why." In fact, the question "why" interrogates the stakes that underlie the representation of animality and asks why we see it as we see it. We will consider this second question in the course of the pages that follow, but without abandoning the preoccupations for the first one, a question that we pose "in the field," which is to say that it permits us to inquire about practices.

The question is vast and the motif is complicated. "Motif" takes on here its double meaning: it is on the one hand the design of a tapestry that we must undo to understand how it was woven; on the other hand, it represents the reasons, the stakes of this elaboration of a particular discourse on animality. The motifs are numerous and intimately interlaced, requiring us to find some of the threads that unravel without breaking. One of them will be, for our work, the one that links the theories, the methodologies, and what we will define as the context of justification of these theories (the loom, the fashion of the colors, the plan of the designer, the integrity of the weave).

THE CONTEXT OF JUSTIFICATION

If the concept of "justification" is perhaps a bit unwieldy, it nevertheless seemed to me to address a good part of what gives birth to a theory and contributes to supporting it. "Justification" here makes reference simultaneously to "just," in the sense of true, because some theories have more of a ring of truth than others; and to "just" in the moral sense of the term, since some theories are in agreement with a certain moral thinking of the world and history;[6] but the term can equally make reference to "justification" in the sense of good reasons, because a theory carries a series of reasons to say and to think things as it says them and thinks them, and since its victory or its stakes can also themselves be explained according to a set of good reasons that are foreign to the truth or falsity of its statements [*énoncés*].[7] This is not a matter only and simply of the social context: it plays an important role of course, initially because it influences our manner of giving a meaning and an interpretation to the relations between organisms, and then because certain theories have a greater chance of being accepted than others that confront beliefs, convictions, and feelings. In this latter case, we witness the combined action of another context, the emotional context: a theory will be perceived as "just" or "unjust" in accordance or disagreement with certain representations to which we are attached.

The context of justification cannot be limited to the social context for still other reasons. The social reductionism that consists in thinking that the discourse on nature is a product of the application of our concept of social structures to the order of nature is a constant temptation when we get mixed up in untying the knots between scientific theories and the contexts that produce them. But this type of approach, as we will have occasion to see, leads to untenable contradictions. Independently of these contradictions, one would have to remark that if the fact of applying social thinking to nature is a form of anthropomorphism, in the reverse one could only maintain that all the forms of anthropomorphism are a simple product of the "social effect" and that they can all be reduced to it. It could be useful to introduce the distinction between two types of anthropomorphic procedures: when ethology tries to find the justification for private property in elaborating a concept like game theory of evolution (when it presents what they call the "bourgeois strategy," showing that the respect for property is a stable strategy permitting the avoidance of conflicts), it presents a form of anthropomorphism that bears the signature of a given society. On the contrary, when it attributes to animals behaviors like the acts of dancing, kissing, or loving—which are identified and named by analogy with human categories—anthropomorphism escapes a social analysis of this type.

The context of justification will therefore repeat the influence of factors that justify the collection of information in one manner rather than another: for example, the fact that a hypothesis influences the *construction* of certain facts and the collection of others. These hypotheses, once adopted, will in many cases justify the theory that presided over their formation. We will speak, then, in terms of beliefs and expectations to explicate this field of context.

It was a matter from then on of trying to apply this grid of constructivist reading to Zahavi's work: not only analyzing his articles but also considering their hypotheses and theoretical propositions in relation to the practices, methods, and contexts of justification in which these theories, hypotheses, and practices emerged and developed. It was a matter of "unweaving" the strands of the context of construction of representations about the babbler, of observing how the expectations of the observer were able to produce the existence of extraordinary birds.

THE PARADIGM OF THE PREDICTIVE EFFECT

Research of this type had already been conducted some years ago, but in a laboratory. Ever since then, the experiment of Robert Rosenthal constitutes a paradigm for the influence of the beliefs and expectations of the observer on the animal observed.[8] The example is a bit caricatured and has been critiqued many times, but it nonetheless remains interesting when one wishes to speak of certain research difficulties in animal or human psychology. These difficulties are tied to the particularity of the object of research: the object is a subject. That is to say that the object is a "someone," who will be irremediably affected by this research. The research then becomes, as Isabelle Stengers says about the Milgram experiment, "productive of existence."[9] As a producer of existence, it partially escapes the experimenter who from this point on remains part of it. We see how Rosenthal's experiment became a producer of existence, and maybe even more so than the Milgram experiment, because it unknowingly multiplied the objects become subjects.

Rosenthal asked his students to continue the research undertaken at Berkeley in the 1940s by Robert Tryon with rats. This research had the goal of measuring the heredity of intelligence. To do this Tryon had tested the rats in the navigation of a maze, selected the most brilliant and the most mediocre, and made sure that they did not reproduce except with partners of the same level. According to Tryon, the learning curves improved for the brilliant rats and worsened for the mediocre ones before arriving, after several generations, at a plateau where the results no longer seemed to vary. The experiment had long finished, but Berkeley had kept some specimens and maintained the process of selection. The descendants of the two lines could thus be put to the test in the maze to be evaluated again. Rosenthal formed several groups of two students and gave to each of these pairs a descendant of the Berkeley rats. He wanted to measure their performance, in other words to see if the intelligent ones were as good or still better and if those who were less so had seen a further diminution in their capacities. The students did their research with, it would seem, all the care necessary, and they confirmed Tryon's hypotheses: the descendants of the intelligent rats obtained better scores in their performances than their idiotic congeners.

The problem was, of course, that Tryon's rats had long

since disappeared from the laboratories and these so-called Berkeley descendants had been purchased for the undertaking and distributed in an aleatory manner into the groups that were supposed to characterize them. Therefore the question of knowing what exactly happened has not to my knowledge received a satisfactory response. We could establish several hypotheses in the mode of a fiction. We could then turn the question around: What would have happened if Rosenthal had asked an anthropologist to take part in the experiment. Would she have perceived the mechanisms that worked to make this experiment the very paradigm of a self-fulfilling prediction?[10]

We will try to consider these hypotheses in some of the terms of the context of justification. In postulating that there was in fact no intent of cheating or lying on the part of the students (which seems to be the case), the anthropologist could have observed how the particular relations were formed between the researchers and the rats (first hypothesis: the relations affect the performances). Imagine that someone places an animal in your hands—more specifically a rat, even if our big white laboratory rats evoke less terror and repulsion than city rats—and tells you that you are holding an animal of superior intelligence: you will not have the same attitude toward it as if they had told you that it was the lowest of cretins. The way of manipulating it, of placing it into favorable experimental conditions, et cetera, will certainly be different from one case to the other. This was one of Rosenthal's hypotheses. The students themselves confirmed this in responding to questionnaires asking them to describe the particular relations they were able to establish with their rats. An intelligent rat inspires confidence and, perhaps more than a rat that is considered to be stupid, appeals to mechanisms of identification that favor empathy. Our anthropologist also may have observed how the rats were held, manipulated, placed, or thrown into the test mazes, calmed before the tests, reinforced after, and so forth.

She also may have seen the way the results were measured, the differential precision of one group from the other (for example, how the chronometer readings were taken), the fact that certain contradictory results were ignored by attributing them to error, and the like. This constitutes another plausible hypothesis: that the data collected are actively selected and filtered.

AUTHORITY AS DISCOURSE THAT RENDERS TRUE

Our anthropologist, an amateur[11] of sociological readings concerning the way science is done, could have been able to read the events in terms of *power,* and pose the question of knowing in terms of one's career or exam results, what the outcome would be for a student researcher who obtained results that contradicted what the experimenter had initially expected to discover. She could also pose the question in terms of *authority,* in the Batesonian sense of the term: how the researchers did everything that was in their power, and in a non-conscious manner, to render Rosenthal's discourse *true,* because it was important to them that it was.

Rosenthal did not seem to foresee all the possible consequences of this nuance between power and authority. It is certainly true that Rosenthal was aware of experiments showing possible bias in human studies. He wrote, "The non-present third party is the principal investigator, who, by what he is, and what he does, and how he does it in his own dyadic relation with the experimenter, indirectly affects the response of the subject who he never encounters. He changes the experimenter's behavior in ways that change the subject's behavior."[12] But when he describes the rat experiments he tells us that he took a series of precautions with the goal of ensuring that each student would be particularly aware that the results of the work would affect neither their grades nor their future career. As soon as it is a matter of grades or of career, we see clearly that what is in play are the elements of a relation of *power* and not of *authority* between Rosenthal and his students. This seems to me to show that the distinction between power and authority is not clearly established.

We should note in passing that the three hypotheses entertained here do not exhaust the entire field of possible hypotheses, nor do they pose the question of the validity of classifications and of the criteria that accompany them. And more than anything, these hypotheses are in no way incompatible—quite the contrary—because they complete one another.

In recalling the Batesonian definition of authority, what comes to mind is that this is doubtless where the true node of the problem resides. What Rosenthal's experiment teaches us is not necessarily located where Rosenthal situates it. The objective of the experiment was achieved because it aimed to test the influence of beliefs on the results of an experiment, but the place or the level where this objective is situated is not exclusively where Rosenthal expects it to be. We can imagine that Rosenthal's students, sensitive to his authority, intuitively understood—perhaps even without being conscious of it—what Rosenthal's real expectations were: that they were wrong, that they were all in error. Definitively, the question that we would have to pose to Rosenthal would be to know on which self-realizing prediction the experiment bears: that of the student regarding the rat or that of Rosenthal regarding the student? It seems that the experiment bears as much on the influence of the students' predictions regarding the intelligence or idiocy of the rats as it does on the influence of Rosenthal's predictions concerning the performances of the students (my students are stupid, or naïve, or will still be deceived by me, Rosenthal). The place of the experiment is no longer in the laboratory, between a stupid (or intelligent) rat and a student, but is rather situated in the reflection of this experiment, between a student defined as someone "who doesn't know" and Rosenthal who defines them as such. To the beliefs of the students in relation to the rats correspond Rosenthal's beliefs in relation to the students when he hopes that they let themselves be duped.

If Rosenthal was not able to disentangle where the biases were created, it is perhaps because he himself became the subject of his experiment and found himself inside the dispositive set up to produce existence. It was not only the rats who became others in this transformative experiment, it was also the students, and above all Rosenthal himself. It is at once both simple and complicated: we can attribute to the interpretation of this experiment two levels of explication that do not—wrongly in my view—receive the same explanatory treatment: given that it is the prediction that makes the rats be seen as intelligent or stupid, one could deduce (in trusting what we know before the experiment) that the rats are not *really* intelligent or stupid. From this perspective, the fact of considering them in terms of the label falls under the realm of error since the rats are not, in "true reality," as they are labeled. But the fact of predicting that the students will be deceived or sensitive to the beliefs or the label does not call into question the definition that Rosenthal attributes to them to predict their behavior. The fact that the student behaves in the manner that Rosenthal labels them will not be interpreted here as an error. The treatment is therefore an unjust treatment: Rosenthal refuses to attribute to his influence on the student the impact that he attributes to the student on the rat. In other words, in Rosenthal's vision, the rat "in true reality" is *not* like we label it—here the label produces "error"—but the student himself, in "true reality," is exactly as Rosenthal labeled him—a deceived student: here only the label predicts— and produces—the truth.

A simple means of getting out of this contradiction between the two explanatory levels is to admit that the experiment is productive of true existence on two separate levels: the rats have really become intelligent or stupid, and the students have really become as Rosenthal expected them to be from the moment that the experiment made them exist as such, on both sides.

EXPERIMENTS THAT PRODUCE EXISTENCE

What would our anthropologist have been able to do with this experiment? How would she have been able to deconstruct a tangle in which she herself was held? In this context, can we imagine that the study of the ties between methodologies and theories can reveal some secrets to us? Can we imagine a context of justification that is revealed? Where we can point a finger and exclaim, "Here, we have now encountered the moment of the subjectivity of the researcher! There is the moment of anthropomorphism! There is the moment of power and interests!"

I do not imagine this is the case for Rosenthal's experiment because it produces existence. The anthropologist could certainly work on the conditions of this production and on its different levels, on the mystifications and beliefs. But contrary to Rosenthal's experiment there is, on the grounds of ethology (except in the novel *Brazzaville Beach* by William Boyd), no mystification, no secret, no unveiling, the ties between the productive conditions of existence cannot receive a straightforward meaning. In simpler terms, I cannot know, based simply on what the animal does and what the author interprets from this doing, which one will produce the existence of the other.

There are ties that can be described or even be created. And there will be hypotheses regarding them that try to give them a sense, a meaning. The areas in which all research modifies its subject, and sometimes simply from the fact of being interested in it, differ in this considerably from a natural world without a soul: such is the case, for example, when Harry Collins and Trevor Pinch show that each course in which children carry out the same experiment together represents, in miniature, the world of scientific research. In inverting the principle of Rosenthal's experiment, since the teacher asks the children to determine without their knowing in advance (by putting a thermometer into a beaker) what the boiling point of water is, we observe that the results will differ by several degrees. But these differences can all fall under a univocal explanation: the thermometer of one student found itself in a bubble of over-heated vapor when they took their measurement, the water in the vessel of another contained impurities, the third student was distracted and let their water boil off and rupture the thermometer, et cetera. It is not possible to proceed so simply for the sciences concerning the study of living beings.

As regards the sciences productive of existence, they are the site of creation of multiple and non-univocal ties between the theories and practices that constitute first their tools, then their results. Beyond this, behind each of the hypotheses that will create ties between a theory, a practice, and the objects to which they are addressed, we can see that there is a more fundamental hypothesis in relation to which we lose all exteriority. And this is where we can begin to observe that the relation between Rosenthal and his students is analogous to that between the students and the rats. This is well-known to sociologists of science who (almost always) take precautions against the most ironic among them: "Our own frame of analysis does not, of course, escape from the same critiques as those that we make of our object of analysis," they would say. But we say this too fast. As if we risked creating an infinite regression. However, if I have properly understood Isabelle Stengers, it is precisely here that we could have started to laugh.[13] We say this quickly as if it didn't have so much importance. As if, between good people, we nonetheless had to take certain precautions against the possibility of a rough-hewn autodidact who could always emerge.

I considered this limit to our investigation essential: the anthropologist who could have worked for Rosenthal would have found herself becoming an element of the experiment, one of its variables, and would not have been able to work with the fundamental hypothesis inside of which she found herself. What the rat became in the particular relation with the student, the student himself became in the particular relation with Rosenthal: the student existed as a producer of intelligent rats. Rosenthal started to exist as a producer of students who produce intelligent rats (or stupid rats to be sure, but we can prefer to look at the bright side of things), in the same way in which Milgram had begun to produce executioners.

What we can see underway in this experiment is a cascade of the production of existences where we find that each of the actors is snared within a hypothesis more fundamental than the hypothesis with which they work. And the anthropologist herself does not escape this rule: in observing how each of the researchers enters into relation with animals, she produces a new phenomenon, a new "researcher–animal" relation that ends up becoming the irreducible third term. The impossible question "How was the animal before the researcher observed it? How is it independent of this observation?" must among other things take account of the fundamental inaccessible hypothesis: "How were they before I started to observe them?" All this is to say that, if it is possible to be constructivist in the lecture hall, the situation becomes especially complicated when one finds oneself an actor on the scientific stage. This, as we will see, is something that the babblers taught me.

Since the fundamental hypothesis is as inaccessible to us as the real in itself, what remains in the field is the descriptive work of "hybrids":[14] animals within the networks of theories, tools, and methodologies; what remains is the work of the creation of ties between *saying* and *doing*. The *sayings* will be, then, in this context, those of the researchers working around Zahavi in a research center in a national reserve in the center of the Negev desert in Israel. These *sayings* build what we call the handicap theory. The *doings* will be understood as the extraordinary activities of these birds who gave birth to this theory of the handicap.

THE DANCE OF THE ARABIAN BABBLER

One of the most surprising behaviors of the babblers is the fact that they dance. The dance of the babbler belongs to that particular category of behaviors for which we cannot say if a good interpretation is provided by means of another behavior. The fact of having called a series of movements "dance" rather than "game" or "rush" already constitutes a particular classification of a regular sequence of actions. This classification will already privilege certain interpretations that are compatible with what we habitually insert into this category and will, above all, exclude others.

The dance of the Arabian babbler is, no doubt, nothing other than a pretext, a tangle in our story. To start with, I remember that the description of the dance intrigued me so much that I had changed my questions to adopt those of our imaginary anthropologist sent to work in Rosenthal's laboratories. I knew that, like her, I could not close off the question of knowing if it was Zahavi's gaze or the behavior of the birds themselves that made the creatures so extraordinary. But I thought that I would be able to unravel some of the motifs of their interaction. It would therefore be necessary for me to join them and watch them coexisting.

The dance is at the center of our story because it crystallizes, as representation, some of the elements of the context of justification. With it appears, for example, the particular dynamic between the facts observed and their classification. Through it, the fact becomes the object of its category: the movement becomes a dance and it is difficult, even impossible, to consider it otherwise, in the same way that the perception of the vase in the figure by Rubin makes it difficult for the naïve observer to subsequently perceive the background as a figure.[15]

Next, thanks to the dance, we will see the emergence of the relations of power between two researchers since this was the place of conflict between two interpretations. In its turn the question will appear—and perhaps in response to this conflict—regarding the influence of observation on the behavior of the birds. This question allows a researcher to reconcile the divergent demands of their work.

Finally, the interpretation of the dance can be put into relation with the whole theory, with the social context, and with the methodologies. The dance of the Arabian babblers constitutes, along with some of the other behaviors of this very particular bird, the point of departure for a simultaneously inventive, original, somewhat reactionary, and above all very controversial theory: the theory of signals and its corollary, the handicap principle.

I am going to try to describe some of the elements of the context for the justification of this theory in recognizing that no interpretation can exhaust its multiple meanings. I gleaned here and there some events, moments where the methodology seemed to imprint a particular mark on the theory, moments where the facts in the field and the social structure in which the discourses describing this field seemed to speak the same language, without my being able to know which gave to the other the means to think as such. But I still have to describe the things—the causes—like the products of the relation between a text and a staging: how the text cannot authorize just any staging, and how the staging gives its own particular meaning to the text.

Before elucidating the handicap principle, we must situate the theoretical framework within which this theory was born as well as the unresolved questions that Zahavi hoped to answer thanks to it.

The handicap principle is a theory that interprets some information from the field in order to then respond to two problematic questions, two paradoxes of natural selection: altruism and the extravagance of signals in selection within species. Consequently, the first context of justification of the handicap theory for us will be, before any methodology or sociology, the theoretical context in which this theory emerges. This context constitutes a frame within which this theory can be inscribed and to which it can be opposed (from which it can overflow). This theoretical context is also one with a justification because it concerns a part of the history of ethology in recent times. At the same time, this history expresses—through the questions it poses and the responses it invents—our beliefs, our utopias, and the way in which we build what defines and constitutes us in relation to animality.

PART I

The Ethological Debates

1. The Theoretical Context
The Two Paradoxes of the Theory of Evolution

Altruism

This first paradox is clear enough and is summarized by E. O. Wilson in the following manner: heroes don't have children. Consequently, how could altruistic behaviors persist in the animal world if they challenged the individual survival and reproductive success of the one who displayed them?

The research of the 1960s placed altruistic behaviors among the most burning questions of natural history (and social psychology). The new concepts of the theory of group selection allowed their explanation, which would in turn further encourage their observation. The social dimensions of behaviors that were generally ignored in favor of the study of structures and functions experienced a return of interest on the part of researchers. Social movements themselves and the research in social psychology seemed to favor solidarity as a privileged domain of research. The murder of Kitty Genovese and the scandal surrounding the inaction of the witnesses were the historical origin of a passel of research on the subject of altruism. Kitty Genovese was that young New Yorker who was killed while she was trying to enter her apartment building one night in 1964. Her cries of pain and her appeals for help moved neither her aggressor nor her neighbors. Safely in their own apartments, the neighbors waited more than thirty minutes before they called the police. During all this time, the young woman was stabbed and beaten without anyone lifting a finger. At least thirty-nine people remained passive in the dark behind their windows. America was shocked when the newspapers published the story. A number of questions emerged at that time and gave rise to a great deal of research on altruism. They started by approaching the problem in its negative dimension: Why didn't those people intervene? Why, in certain circumstances, are we not altruistic? What are the phenomena that justify inaction?

The murder and the scandal of the witnesses' inertia only constitute, in my view, the launching event of this new research. An optimistic vision of humanity and the world seems to me to favor the study of some particular behavioral fields. These new domains turned out to be not only exciting, but equally useful since campaigns of civic education could be put in place on the basis of research in social psychology: in effect, as soon as we know what inhibits the pro-social or altruistic behaviors, we would be able to act on these inhibitions (here it is necessary to look at the work of Erwin Staub on this subject).[1] Parallel to this research with an educational focus—and this is not a coincidence—interest started to grow, under the same optic, in the learning of violence.

A very singular definition of the human emerges from a small number of simple statements derived from the research program in social psychology: pro-social behaviors are not exhibited because they are—beyond a certain age or due to social circumstances—inhibited; aggressive behaviors are exhibited because they are learned, particularly from the media. The behaviorist doctrines issuing from Skinnerian theory are certainly not foreign to this great utopia of learning and unlearning of social roles. But what especially comes to the fore from all this is the new binary anthropology that defines pro-social behaviors as intrinsically present—nearly innate, which means we can only "unlearn" them—and which defines aggressive behaviors as acquired—since one must learn them to demonstrate them. Behind this anthropology is a vision of a pacific and unified world that seems to color the theories and utopias that underlie it.

During this same time period, ethology also began to show the influence of this new anthropology and cleared the way for a vision that instead situated organisms and individuals in a network of exchange and even solidarity rather than beholden to a worldview predicated only on the individual struggle for life. The year 1962 saw the publication of V. C. Wynne-Edwards's *Animal Dispersion in Relation to Social Behavior,* and the following year the first edition of Konrad Lorenz's *Aggression: A Natural History of Evil.* Despite numerous differences and, it would seem, independent sources, the problems raised by both these authors and the new responses they made to them are remarkably similar.

Wynne-Edwards postulates that groups are ceaselessly confronted with a problem of the regulation of demographic density. If we believe the Malthusian law as Darwin applied it to the animal world, demographic growth follows a spike in the curve related to the growth of resources. Animal populations must therefore sooner or later find themselves confronted by problems of the overexploitation and thus exhaustion of resources. According to the traditional conception inherited from Darwin, the only factors of demographic regulation would be due to the pressure of external elements such as the climate, increase in predation, parasites, and the destruction of resources, or again the effect of continual competition. Now, as Wynne-Edwards states, we observe that populations always vary in much stricter limits than these traditional accounts would allow—much less predict or explain. When food resources are composed of living organisms (animal or plant), there are, according to Wynne-Edwards, "security barriers" or active mechanisms used by individuals themselves that prevent their overexploitation. Wynne-Edwards uses the metaphor of the management of capital: we could consider the prey as the interest on a perpetual capital; as long as the consumer only makes use of the interest, the capital remains intact and is then able to produce new interest in the following year. The principle of demographic regulation therefore resides in the fact of leaving the capital intact. However, this principle cannot continue to function properly unless the populations restrain their numbers before it is too late, that is to say before external constraints—the diminution of the capital—eliminates the entire group. Animal populations adopt systems of convention and substitutes: territory and hierarchy. From the moment when the possession of a territory—or a rank in the hierarchy—is the necessary condition for the reproduction of power, each of the members submitting to this convention collaborates in the homeostatic regulation of demographic density. The convention excludes the right to reproduce an excess of individuals. The "territory" convention is thus the material substitute for resources (the food and the females to which it, other things being equal, grants access): "the essence of territoriality," Wynne-Edwards writes, "is therefore to regulate the number of winners and to divide the population between 'the haves" and 'the have nots.'"[2] Similarly, if the "consumers" are programmed to defend larger territories than what real need requires, the rule of the program—"If you can, defend a large territory"—preserves the integrity of the capital in preventing overexploitation. From this perspective, not only would territorial behavior be an adaptive convention, but a *social* behavior that benefits an entire group. In refusing the right to feed or reproduce subordinate individuals, the "hierarchy" convention performs precisely the same functions of resource preservation as the convention of territory. In this respect, it is also an abstract and conventional substitute for these same resources.

Here we can see a new definition of sociality taking shape, independent of gregariousness: the social animal is the one that agrees, through the system of conventions, to restrict its individual interests for the benefit of the group and the species.

New concepts that enable giving meaning to these observations, explaining them and organizing them, build up around this new paradigm and will give birth to the theory of group selection.

THE THEORY OF GROUP SELECTION

To the question of the benefits of altruism, the theory of group selection responds that groups within which solidarity prevails will experience a better fate than competitive groups. Competition, in this perspective, is only considered between rival groups. A group where the members do not respect the conventions will be a group at the mercy of permanent, violent conflicts, or will lead inexorably to the loss of the individuals that make it up since it will destroy the capital of available resources.

From this perspective, all behaviors whose end goal is the well-being or survival of the group will be well-adapted behaviors.

Lorenz arrived at similar conclusions departing from a very different question: in what way can this evil that is aggression be a good? If aggression plays a well-understood role in predator–prey relationships, we can pose the question as to its persistence in intra-specific relations.

According to Lorenz, the primary function of aggression would be to divide and spread out individuals into a larger space with the goal of avoiding the overexploitation of resources. Lorenz's metaphor, too, is singularly derived from the domain of liberal economy: if in a certain region, a certain number of doctors and bakers want to find their way of making a living, it would be good to set up their practices as far away from each other as possible. This same logic can be applied to animals occupying a given space: it will be most functional if they divide and spread themselves out as evenly as possible within the available vital space. The danger of having one animal species excessively populating a territory and exhausting all the food resources in a part of the available biotope is eliminated in a simple fashion through aggression.

Aggression therefore plays the role, in Lorenz's hypothesis, of the regulation of distance and the distribution of individuals within space. This distribution comes down to territoriality.

We can underscore here the similarity between Lorenz's arguments and those we mentioned in Wynne-Edwards. But once he sets them out, Lorenz becomes concerned instead with understanding and showing the existence of mechanisms that will have the function of channeling this aggression, of "civilizing" it. These mechanisms will be the mechanisms that preside over the formation of ties.

Initially, two particularities can be drawn from Lorenzian anthropology: first, if it is true that aggression is a "necessary evil" since it is an inherent necessity to all social life, this same necessary evil is the condition for social ties. Without hate, no love is possible.

Within Lorenz's anthropology we will be shocked to discover the same unaltered pre-Darwinian arguments that attempted to discern the reasons for why God would have invented evil. Divine benevolence could be uncovered at work within grand designs whose mystery and impenetrability authorized the most fantastical creativity among naturalists: in this way one could put forth the problematic concerning the torture inflicted on the caterpillar when the larvae of the ichneumon wasp begins to devour it from the inside—taking care not to damage the essential organs so that she will remain alive. Lyell, careful to decipher the moral content in nature—and thus to prove God right—considers ichneumon wasps as a fortunate limit imposed by divine benevolence on the voracity of caterpillars. Without it, caterpillars would have already destroyed human agriculture.[3]

Lorenz's utilitarian argument—the idea that territory fulfills a function, is useful—would no doubt with analysis reveal unexpected stakes. Effectively, in this perspective just as in that of Wynne-Edwards, we see the application of the notion that territory is in some way a necessity and not, according to the words of Gilles Deleuze and Félix Guattari, an emergence.[4] The logical consequence of this account—aggression precedes and establishes territory; the requirement of not overexploiting is in some sense the "cause" of territory— finally claims that territory is a necessity for the survival of the species. The logic of this reasoning and what it moves toward establishing ends up creating a singular link between ecological preoccupations, data coming from nature, and sociopolitical arguments. We will come back to this.

Next, another particularity of Lorenzian anthropology: aggression is not only a necessity inherent to social life and the condition of social ties (in other words, an evil for a higher good), but beyond that, it seldom leads to death. These claims are based on observations of certain behaviors—above all rituals—whose function was, according to Lorenz, to inhibit and reorient intra-specific aggression. Death, as Martin Daly remarked, thus disappeared from Lorenzian analysis and became invisible for his readers. And when, at the outbreak of a conflict, an unfortunate individual nonetheless died, the causes for this exception were attributed to accident or pathology.

The vision of the world as a morality tale thus influenced the interpretation of behavioral phenomena and even the manner of recognizing their existence, and therefore of assembling them.

After a period of enthusiasm, the theory of group selection showed signs of weakness. Many critiques were made, most notably regarding the fact that the theory could not explain why altruistic populations were not invaded by cheaters. Before explaining this critique of the contents, we could perhaps consider, in a more externalist way, that the theory became more fragile in the face of criticisms because the times changed and social movements no longer sustained the same type of discourse.

RECURSIVITY OF CONTEXTS OF JUSTIFICATION

The critiques addressed to the theory of group selection's ideology should have, in principle, indicated to us the meanings of these changes. This did not seem to turn out to be the case. These critiques brought us nothing but a sense of confusion and insoluble contradictions that block establishing a context of justification that would have even slightly more coherence. This theory of group selection that sees the natural world as a moral world where solidarity prevails will be experienced (for some writers like Stephen Jay Gould[5] or as I myself suggested on the previous page) as the resurgence of the tradition of natural theologies. A quick parallel with other analyses[6] based on some of the texts by the anarchist Pierre-Alexandre Kropotkin[7] could lead us to think that Lorenz's theory of peace and solidarity displays some affinity with egalitarian and revolutionary ideas. This is not the case, even if the message that the theory of group selection propagated in the literature and the popular media[8] presents clear characteristics of generosity: the vision of a pacific world of cooperation and solidarity in which each of the individuals is open to the good of the others. Here the most emotional context of justification was able to operate: the story that recounts the theory of group selection in this particular version is a "beautiful," edifying story.[9] However, a first detour is already made here: from the common good, we have passed to the good of others, without distinction. An attentive rereading of Wynne-Edwards (especially the citations referred to above) shows what remains more implicit in Lorenz: the common good goes by way of *certain* others (those who have) to the detriment of numerous others (those who have not).

It is without doubt this nuance that led some more suspicious historians to see the ideology of the 1960s as an ideology that favored the idea that social behaviors must necessarily be pro-social behaviors, which is to say socially adapted behaviors that do not benefit the individual, but the group. Behind this type of vision rests the belief that there is no conflict between the well-being of the group and the well-being of the individual: the theory of group selection subsequently appeared, according to this analysis as described in the words of Helena

Cronin, "as a crude apology in favor of capitalism. This apology is reinforced by the belief that if there is a conflict between the interests of the group and those of the individual, those of the group will come first. The ecological vision becomes a holistic vision considering the world as a super-organism where everything is adapted."[10]

A social context of justification seems very difficult to outline: to begin with, it now plays a role in the production of critiques, obliging us to undertake a regressive work—it would be necessary for us to analyze the context of justification of the critiques addressed to the theory of group selection to evaluate the stakes of their disputes;[11] then again it could have operated in a plurivocal manner to favor the success of the theory.

The interpretation of the success and failure of a theory is therefore itself the site of complex stakes. These contradictions must not discourage us, however. We can approach the theory of group selection from another direction that seems to suffer less from this Russian doll effect: beyond and beneath this social context of justification in mosaic, a powerful emotional context will play a non-negligible role in the attraction that this theory can produce. When I give seminars and conferences on the subject of animal altruism, I can clearly identify the emotions of the audience at the exposition of one or the other theory: the sociobiological theory gives rise to astonishment—and sometimes to rejection, if it is described in the caricatural manner that its own authors use and abuse; Zahavi's theory gives rise to a joyful laughter (before then eliciting surprise, absurdity, and paradox, astonishment, and then we begin to see how the unfolding of the process of identification happens); the theory of group selection gives rise, in a very univocal manner, to sympathy. These are what I call "pilo-erective" stories or ideas because the emotions they give rise to are of a very particular nature. Stories of dolphins that save their congeners, of lemmings that disappear for the survival of others, of apes who adopt orphans, these are all edifying stories that "turn out well." We will now return and consider a much vaster approach to how moral fictions or morality tales function in our way of reconstructing the world.

LIMITS TO THE THEORY OF GROUP SELECTION

Now we will return to the critiques that were made about the content of the theory. The primary argument, as I indicated above, points out that the well-being of the group cannot prevent the intrusion—and the invasion—of a mutant strategy that consists of exploiting others. A mutant individual will be advantaged and will rapidly supplant the group, leaving in its wake a myriad of little egoists that are themselves cheaters and thus advantaged. In sum, the theory cannot totally resolve the paradox.

Consider one of the examples of deliberate behaviors permitting, according to Wynne-Edwards, animals to voluntarily regulate their demographic density. Starlings, who reunite in the trees at dusk, make of these reunions opportunities for the exchange of information, notably on the size of the group. Wynne-Edwards qualifies these behaviors of information exchange as "epideictic displays." These exchanges of information are generally followed by a regulation of the population. Depending on the species, one can bear witness to two different strategies: in the first case the hormonal system is said to be sensitive, in one respect or another, to the stimulations produced during these reunions, which has a regulating effect on the reproductive process during the following season. These stimulations can be auditory stimulations. The hormonal system of each of the individuals would thus be influenced by the noise that the other members of the group make. If there is a lot of noise—the population is numerous—the reproductive system will be partially inhibited. This is the strategy adopted by the starlings. In another case the group can separate and the part of the group that is excessive in relation to the amount of available resources will migrate and establish itself elsewhere. From this perspective, as we have emphasized elsewhere, species respect a sort of social convention permitting them to escape extinction (due to the overexploitation of resources) through restraining their density. We should mention in passing that this sort of migration was already the solution put forth by Kropotkin regarding beavers. In his eyes this migratory pattern amounted to clear proof of the falsity of the Malthusian theory while simultaneously being a motor for evolution.

This system of solidarity cannot, however, remain stable: imagine a mutant strategy, an individual whose program will stipulate not to follow the rule "cry honestly," for example a starling that would adopt what John Krebs[12] called the beaugeste strategy (a reference to a tactic used by a unit of the Foreign Legion that, with a loud backup of noise, deceived its enemy as to their real numbers). We could at the same time consider that an individual was sensitive to stimulations emitted by others and not by their own, and that in crying out the individual "cancels out" the effect the cries of others have on the individual's own system (which is altogether plausible, a bit like how children will sing at full volume in order not to hear you): in this way it inhibits the reproductive system of its congeners and will therefore have many more descendants than the others who cry less loudly. If these descendants inherit the parental talents and cry just as loudly, they will rapidly invade the population.

Note that in the long term, if all cry out loudly, there are at least two consequences possible: either the group disappears, or the equilibrium of the strategies is modified. The first case would lead to the fact that, in principle, the noisy groups will attract predators and disappear. Nevertheless this possibility is not certain: Darwin was surprised that some animals were so noisy but survived in spite of it, and Zahavi[13] offers an adaptive explanation for the act of crying out loudly. Loud calling could thus not be contrary to survival. Our first possible consequence is not therefore a necessary one. From there we fall into the second branch of our alternative in terms of possible consequences: the level of sensitivity (and of inhibition) rises with the intensity of the cries, and each one therefore becomes less sensitive to the cries of the others. It is thus perfectly possible to "cheat" and the theory of group selection alone cannot fully account for the mechanisms that prevent a mutant strategy from invading altruistic populations. We end up at best, in cases of the emergence of mutant strategies, with undecided equilibria between two strategies (cheat–cooperate). The problem of the existence of altruistic behaviors thus remains in its entirety.

Nonetheless, as some theorists like William Hamilton[14] remarked, there exists a condition that permits the system to be stable: it suffices that altruism be reserved to relatives. In favoring the members nearby, the altruist also favors his own genes, present in many of the members of the group.

THE SOCIOBIOLOGICAL THEORY

In 1964, while observing bees, Hamilton succeeded in applying a theoretical tool that allowed the resolution of the paradox of altruism for the insect sciences—at least initially. Without entering into the details and the important technical considerations, we could summarize the problem—and the solution—in remarking that Hymenoptera show certain genetic characteristics (haplodiploidy). This characteristic has the consequence that the bees are genetically closer to their sisters than to an eventual descendant. A bee who dedicates herself to her sisters rather than her eventual descendants favors, by her gesture, the transmission of her genetic information. The fundamental tenet of the theory is that altruistic behavior is genetically determined. If the individual favors the carriers of copies of its genetic baggage, the "nepotistic" altruist assures the maintenance of altruism. Genetic altruism, rather than disappearing in the act of sacrifice, will therefore multiply the chances of being genetically present in the next generation.

The model of kin selection—of what will become, with E. O. Wilson, sociobiology—was also formalized by Hamilton and substituted for the theory of group selection. A series of empirical observations that seemed to confirm the theory and even extend its field of application responded to this theoretical condition. The phenomenon was not only present among the Hymenoptera, but could be found in other species despite the absence, in these others, of that characteristic of haplodiploidy.

Among birds, according to the sociobiologists, individuals intervene in an active manner in the reproductive success of their related congeners; we have already spoken about the role of the helpers at the nest. In this way, we were able to say that altruism was a nepotism. Indeed, in certain species of birds it seemed that alloparental behaviors followed very precise rules. Stephen Emlen describes them along these lines, I paraphrase: *1. Help only if the potential recipient is a member of the clan in which you live. 2. If you are not related to members of the clan, do not help since you did not have your own children who would in their turn have children. 3. Help the closest relative among those who reproduce.*[15] These rules give meaning to the regularity of observations and actively construct the frames within which the relations observed can be interpreted. Numerous examples were related, which showed that the help was generally correlated and could even diminish with the degree of relatedness. This would be the case with numerous birds who are "helpers at the nest" like the white-fronted bee-eater, pied kingfisher, black crow, Galápagos mockingbird, and the hoatzin.[16]

The figures are not however always clear and univocal: if the observations of the Florida blue jay led some authors[17] to think that help is oriented toward a relative, others conclude, regarding the same species,[18] that it is not keyed to the degree of relatedness, but rather provided haphazardly under the influence of the demonstration of releasing stimuli: help will be oriented then not toward family, but toward the spatially close individuals.

This last example of the blue jays first of all raises the

question of the reliability of the collection of data. Next, the difference between the manner in which it is distributed in the group leads to asking what really covers kin selection. We thus come to one of the critiques that was addressed to sociobiology: help is not oriented toward relatives, but toward those who are in proximity to the bird. Generally, as these are relatives, the maintenance of behavior will be assured. Some sociobiologists themselves think that the best manner of transmitting genes is in the end to occupy oneself with one's own children.[19] Help would not in fact be founded on—and the most subtle sociobiologists work with this hypothesis— the mechanisms of recognition of relatedness, but on spatial rules and criteria of proximity. This is not, however, as we have seen, the account of sociobiologists like Stephen Emlen[20] or Bruce Waldman.[21] The latter dedicates the majority of his research to separating birth-related animals from one another in order to evaluate their capacities to recognize one another when they are reunited, redeploying for his readers the marvelous fiction of the beauty mark from the most poignant novels of the nineteenth century.

CRITIQUES OF THE THEORY

The critiques that reproached the implausibility of sociobiology were frequently situated at the level of *proximal causes* that sociobiology attributed to the power of *final causes*. When we speak about the origin of a behavior or a structure, we generally refer to an explanatory system invoking the causes of this behavior or of this structure. Of the two series of causes generally invoked, the first takes on the task of explaining the *how*—it is the series of proximal causes or close causal mechanisms. The other takes on the task of understanding the *why* of the structure of behavior, which is to say analyzing the functions by considering them as conditions for the maintenance of behavior. This series of causes is called final causes because it explains the maintenance of the behavior or structure by the goal that they fulfill, which is to say by the fact that it playcd an essential role in the survival of those who possessed or possess that behavior or structure. In other words, the proximate causes take account of the mechanisms that will release the behavior: the motivation of the individual, the stimuli and stimuli signals, the antecedent causes. The final causes explain, for their part, the functions of behavior, which is to say the reasons for its maintenance in the course of evolution.

If we say that the animal chooses to help eight cousins because they share with it an eighth of its variable genetic baggage, or because it can, by its gesture, assure the maintenance of its genes thanks to the replicators that it saves, we make an error of explanatory levels. This error consists in using an explanation at the level of (motivational or releasing) *proximal causes* that instead applies to *final causes*. In other words, the error is that of confusing the causes and effects of behavior. This cannot be understood unless we know that, in a temporal and historical process like evolution, effects play, in a second stage, the role of causes, in the same way as in the homeostatic models. Whenever an individual orients his aid toward a relative—and this help could be for a number of reasons, sometimes simply because they are nearby—it allows for the maintenance of the altruism genes in the population. The particular manner of orienting aid is therefore under the sway of final causes. Thinking that an animal makes knowing and complicated calculations to assure the best duplication of its genetic baggage amounts to attributing to proximal causes what instead belongs under the heading of final causes. The confusion between proximal causes (the individual helps the other because it receives releasing signals) and final causes (the individuals who orient their aid toward close relatives allow for the maintenance of altruism) frequently constituted the foundation for critiques addressed to sociobiology, the critiques attributed to the proximal level what the sociobiologists situated at the final level. It even happens that some sociobiologists themselves saw the choice of orientation of aid in terms of knowing calculations (I must save at least two brothers or eight cousins) at the level of proximal causes motivating the aid behavior. The most radical, like J. Philippe Rushton[22] or even Bruce Waldman, in postulating mechanisms for the recognition of genetic relatedness, often fall under the blow of this critique. According to John Maynard Smith,[23] we have too frequently accorded an overly important place to kin selection in obscuring the fact that cohabitation between congeners is frequently little more than a modus vivendi between competing individuals. Another argument of the criticisms was—and, I think, is always—the splitting of behavior or the organism into genetically determined adapted traits that are relatively independent from one another since they were shaped by evolution. This division was in fact the object of two different series of critiques: first, an epistemological critique, and following that, a political critique.

The epistemological critique addressed two major objections to sociobiology: the panglossianism of its explanations and the reification of the gene. Panglossianism is a form of pan-selectionism.[24] The concept of pan-selectionism, panglossianism, or even hyper-adaptationism came out of utilitarian thinking when it was pushed to the extreme: pan-selectionism is the tendency to think that everything that was selected was chosen because it represented an adaptation. We now understand why this pan-selectionism is also characterized as panglossianism: it brings up numerous analogies with the Voltairean heroes for whom the "metaphysico-theologico-cosmolo-nigological" doctrine holds that "everything is done for an end." From this perspective, it follows that everything that exists is useful to a function and is adapted.

When the critiques denounce the reification of the gene, they highlight an important drift: the gene, which was not initially applied to behaviors except as a hypothesis, acquires the status of thing–cause in lieu of the presumed factor. It thus passes from the level of inference to that of established fact.[25]

This objection is inscribed in a more general critique of

the reduction of behavior to the gene. It scrutinizes the confusion of specific behavioral traits with the general possibilities of behavior,[26] a symptomatic confusion that Gould denounces as a consequence of bad habits taken on by Western scientific thinking: atomism, reductionism, determinism.

Next, the other series of critiques, this time of a political order, criticize sociobiology for its conservatism, the legitimation of practices that are racist, xenophobic, eugenic, sexist, and so forth. The literature on this subject is too vast to be cited. As the critique of the theory of group selection has already suggested to us, one has to be very naïve to think that the social context of justification can be, in a direct manner, translated on the basis of the political critique. It would be necessary to suppose for example that science is the locus of collaboration of the extreme right and dirty tricks. Some critiques suggested this, but we will not revisit this discussion. We can however underline an interesting characteristic of this objection addressed to sociobiological scientists: the "political" criticism of the conspiracy presents the same symptom as the sociobiological reasoning itself. We just mentioned the fact that sociobiology was criticized from an epistemological point of view for its hyper-adaptationism, which is to say a radically utilitarian vision of behaviors and organs. Now, arguing, as sociobiology does, that "if it's there, it's because it is useful," is to pretend that the current function takes account of the origin of the behavior. Hyper-adaptationism thus reflects a confusion between genesis and utility. The critiques that see sociobiology as a conspiracy make this same confusion between genesis and utility since they imagine that science is constructed through a precise outline, and that the ultimate consequences of its discourse were, from its origin, the causes of its elaboration. Gould's concept of "exaptation,"[27] that dissociates primary utility and secondary utility in the genesis of organs and behaviors, was able to find a wide field of application in the domain of the external critique of discursive complexes.[28]

SOCIOBIOLOGY AND THE CONTEXT OF JUSTIFICATION

Beyond all the political objections that one could make of it, there is, I think, a sociobiological stake that must play an important role in the context of justification: it is the stake of reductive determinism. It plays the game of the confiscator. Confiscator of liberties, some would say, but that is not, in my view, the real stake of confiscation. It is also a confiscator of responsibilities, since with genetic determinism, bad faith in the Sartrean sense (it is not me, it is my genes) finds a naturalistic foundation.[29] But that is still not the true source of the problem even if this confiscation of responsibilities would be the corollary of it. The real confiscation is that of competencies: attributing our most private and most relational social behavior to causes inaccessible to anyone but specialized technicians. This is to take away from the relational space and the resources of care and management everything that could add up to social behavior. In my view, this is where the true danger of genetic determinism resides: it confiscates, for the benefit of the technicians, the social competencies of each and every one of us to participate in the intimacy of the suffering and difficulties of the relational life of the other. Sociobiology, as a discourse that justifies the attribution of competencies to some and legitimizes the disempowerment of others, found a special place in Western countries—especially the United States.

Sociobiology provoked a general outcry. There we can begin to glimpse the clues for our context of justification. What assured sociobiology such a success was its extraordinary mobilizing capability. Seldom have books of such virulence been written against a theory—a theory that, it is true, often significantly overflows the frame of the established facts in order to launch into very audacious speculations; and some defenders, such as the sociobiologists David P. Barash and J. Philippe Rushton, put a particular zeal into offering up phrases like "nature is sexist" and the like.

This mobilizing force made sociobiology the perfect place

for denunciation. If we weigh the texts that came out after the publication of Wilson's *Human Nature,* we notice that sociobiology was not only the site of a denunciation in search of objects, but equally the opportunity for a clarifying expression of ideological positions. It offered a somewhat more consistent adversary than the windmills against which the democratic melancholics battled.[30] Sociobiology seemed to exert its mobilizing force on the two camps of its detractors and its partisans: it carried a detestable political message—and clearly detestable for some—and thus gave them a locus of denunciation. But for others, it also was the harbinger of a discourse that compensated for what was in the process of being lost; this is certainly the case, and we mustn't forget that sociobiology arrived on the scientific scene at a moment when the emerging economic crisis undermined the illusions created by several decades of expansion. It therefore appeared to provide, at just the right moment, the justification for unsurpassable constraints: *Le Monde diplomatique* chose what is in this context a revealing title: *Science in the Time of Austerity.*[31] Family values also seemed to be strongly supported by this redeployment of naturalism or, as Michel Maffesoli called it, "this effervescence of social vitalism."[32]

But it is not a matter of pulling the rug out from under all external critique in claiming that sociobiology owes its success to its detractors rather than to its partisans. Nor is it a question of denying the relevance of an analysis that presents the crisis of values or the crisis overall as a propitious context for the reception of these theories. It is a question simply of relativizing this reception by simultaneously taking account of the mobilizing force supporting this theory and its reception and its capacity to stimulate passions around itself. In a certain way—notably as site of denunciation—sociobiological theories play the role of theoretical scapegoat. Things seem to have calmed down after some years and an epistemological debate seems to have been substituted for the political debate—which has not prevented Rushton and others from continuing to write and publish what sometimes amount to dangerous stupidities.

Let us return to these epistemological critiques. One among them, rather than calling the model into question, imposes limits on it. Recall the blue jay studied by Woolfenden and Fitzpatrick: if the blue jays do not aid their relatives, how can the aid be maintained? There is a need for an alternative explanatory paradigm that can take into account these types of facts. Astonishingly enough, this alternative paradigm does not take shape in the field, nor under the microscopes tracking sequences of DNA, but in the places that appeared a priori to be the furthest from the empirical process: the computers used in the simulation laboratories. One tool derived from the game theory of evolution allows us to consider—and even prove this to be the case since it is a mathematical tool—that in some circumstances, natural selection can favor a system of reciprocal altruism and maintain it in a stable fashion within the population. Altruism thus became a strategy defined as ESS [evolutionarily stable strategy], or as a stable strategy from the point of view of evolution. This model gave rise to the theory of *tit for tat* or reciprocity.[33] The observations of some researchers came to confirm the predictive value of this model. The real world seemed to conform to what the virtual world in the computer had created: that altruistic animals really appeared to use strategies based on reciprocity. And it was those strategies that allowed solidarity to be maintained.

THE THEORY OF RECIPROCITY

Among the Galápagos finches, some of the birds orient their aid toward close relatives while others seem not to do so. Those from the latter group preferentially aid couples where one had initially helped in feeding them before as helper-at-the-nest when they were chicks.[34] Among the stripe-backed wren,[35] a helper who manages to reproduce will receive aid from those whom they have tended or fed. Among the white-fronted bee-eater and the Mexican jay,[36] the birds alternate roles of parent and helper and thus find themselves inside a system for the exchange of aid. The ethologist couple J. and S. Ligon[37] observed that among the African wood hoopoe, the parents generally allow the children who reach maturity to remain at the nest. They then play a role of helper for the subsequent broods. When a territory is opened up, one of them may move there with the help of the young he helped to raise and who will move in with him.

Vampire bats[38] regularly practice exchange of food in the mode of reciprocity: the fact of succeeding at feeding oneself daily is not dependent on individual competencies, but in reality rests on chance. If an unfortunate vampire bat finds itself in danger, she will benefit from the sharing that one of her companions will provide in regurgitating a bit of her meal. Later, she can, in her turn, return the favor to the one who fed her. It would seem that in this case, reciprocity is assured by social mechanisms that penalize cheaters and exclude them from the social network.

This model was, however, called into question on the basis of empirical data: the mechanisms of penalization and exclusion necessary to the stability of the system seem often to be absent—or at least invisible to the eyes of the researchers. If this is the case, the paradox remains open to hypotheses. Zahavi's handicap theory—elaborated at the start of the observation of altruism among the babblers—offers to resolve the paradox. However, its ambition doesn't stop there. Indeed another great paradox of natural selection remains: the "extravagance" of signals in natural selection.

It is here that the great originality and richness of the handicap theory resides: with exactly the same hypothesis, it aims to resolve the second paradox at the same time as the first one, and considers that altruism and the extravagance of signals used in the quest for a partner are in the end two facets of the same problem. But before turning our attention to the resolution of these paradoxes, we will briefly study in greater detail this second facet mentioned: What is the function of the feathers, colors, and ornaments that are above all somewhat encumbering? How can they be a means of intimidating rivals while also being considered as a form of attraction in the eyes of females? These are the questions that, following Darwin, have been addressed by theories of sexual selection.

Sexual Selection

The extravagance of the feathers of the Argus pheasant,[39] the spread-out tail of the peacock, or the antlers of the Irish elk[40] gave the theories of sexual selection as much of a runaround as the extravagant behaviors of altruism brought to natural selection. In 1871 Darwin put forward the hypothesis of a selective pressure different from that of natural selection: sexual selection.[41] This selective pressure operates at two levels: that of proximal causes with the hypothesis of the "good taste" of females in the choice of attractive males, and that of final causes by which these qualities privileged by the females will spread through the population since the males that possess them will have greater chance to reproduce. The arbitrariness or the extravagance of some of these characteristics and the waste they imply was of little concern for Darwin: they increase the risks in the face of predation, rely heavily on an expenditure of energy, and can constitute a real burden for those who carry them.

Darwin is certainly not a hyper-adaptationist: sometimes he even attempts to seek out the imperfections rather than the adaptations in nature since these represent a double interest for him: from the methodological point of view, imperfections provide a means for him to demonstrate the gradual character of evolutive processes; from an ideological point of view, imperfections are the proof Darwin requires for eliminating the role of the Grand Architect from his explanatory system. We recall under this heading the extract justifying the behaviors of the cuckoo bird or the ichneumon wasp. Perfect adaptation was the major argument of the utilitarian creationists for proving the existence of a divine plan. Darwin, in his care to refute these cumbersome schemes, will sometimes focus his attention on the imperfections to the detriment of adaptations.[42]

The interpretation of the arbitrariness of secondary sexual characteristics, such as it was conceived by Darwin, was revisited in 1915 by Ronald Fisher and strongly influenced later theories: the characteristics maintain themselves because they spread through the population according to the processes of self-reinforcement. A characteristic assures a female that her own sons will be chosen in their turn if they inherit it. The sought-after quality is thus reinforced by the choice of the females and loses, by this fact, its arbitrary quality.

Alfred Russel Wallace, in 1889, proposed an alternative hypothesis to that of the "good taste" of the females: what we could call the theory of the "good sense" of the females. According to Wallace, beauty and quality tend to coincide: the female does not choose the male for his ornaments, but for the noticeable qualities that accompany them and that constitute the external privileged application of maturity and vigor. The arrangements of colors are the consequences of physiological phenomena. They generally coincide with structures and change at the points where the functions change, also revealing the structures. These are thus the logical and not the aesthetic reasons that determine the attraction of females to these qualities. In line with the Wallacean theory of the "good sense" of females, numerous theories over the past few decades have tried to understand the ties between the extravagance of certain male characteristics and what predisposes females to choose them.[43] But after explaining the preceding considerations, can we still say that extravagance is a paradox? Darwin and Wallace do not give the impression that it posed them such a great problem, and each of them offers a solution that seems to answer the question.

How did the insistence on the paradoxical aspect of the characteristics of intra-species extravagance come about?

EXTRAVAGANCE AND THE MORAL OF
NATURAL HISTORY

We could entertain the hypothesis that the act of defining some characteristics as extravagant corresponds to a certain vision of nature according to which all that is, has good reasons to be. Extravagance as an arbitrary characteristic is in some ways at odds with the specific adaptation of each element to the entire system. This corresponds to our double tendency, to our double belief: first, thinking that things must be linked to a particular meaning. In this way, the big feathers must signify something—not be "free," in sum—and must not be, as some authors suggest, a simple lateral consequence of the growth of the entire body.

As for the other belief, it leads us to construe the guiding principles of nature as moral principles: even if generous, this nature must be economical. It is wise, and therefore neither eccentric nor extravagant. Panglossianism is built from a series of stories where the moral agrees with our natural judgments: think of our love of stories that end well, and think too of what Roland Barthes teaches us about our mythologies when he describes, as a staging, the wrestling match in the course of which a flagrant wrong sees the carrying out, in a spectacular manner, of its reparation.[44] Think also about that belief in the justice of the world[45] that is summed up in that well-known adage "Not everyone gets what they deserve and not everyone deserves what they get."

The law of might makes right, and perhaps more particularly its intellectual variant for which "David and Goliath" constitutes the paradigm—the law of might makes right becoming here the law of the more "cunning"—is, in the end, a moral law that structures our mental and moral universes.

Extravagance, from this perspective, must either have a meaning or be irrevocably counter-selected. The moral fiction is translated into the fact that extravagance has an adaptive function, a good reason to have been selected (i.e., an excuse for still being among us). In adaptive terms, the justification of extravagance can be seen as playing the role of rationalizing our mythologies.

Gould wonderfully illustrates our tendency to make a moral story out of natural history: the Irish elk, extinct now for close to ten thousand years, had enormous antlers.[46] The story of its extinction—in the same way as that of the dinosaurs— gives way to moral fictions that seem to exclude the extravagance so contrary to the wisdom of nature: the traditional histories tell us for example that what, at first, must have been an advantage, experienced a growth such that the antlers instead became a mortal burden: "Like the sorcerer's apprentice the giant deer discovered only too late that even good things have their limits. Bowed by the weight of their cranial excrescences, caught in the trees or mired in the ponds, they died."[47] The deer had thus—like the dinosaur or the marsupials of the New World—committed a "fault." Extravagance, as its name indicated, is an error or better still, according to Lorenz, a "stupid product of intra-species selection."[48]

The question of extravagance takes on a different meaning according to whether we consider the antlers of the elk—just as the antlers of normal deer—in the frame of natural selection, where they play the role of an arm of combat, or within the framework of sexual selection, where they take on the role of an ornament of attraction.

For the Darwinians of the nineteenth century, the success of the individual is measured principally by the number of battles won and enemies destroyed: in this frame of reference, the antlers are undoubtedly a weapon against predators and rivals. Darwin, in *Descent,* nonetheless considers the hypothesis according to which the antlers could have the function of attracting females. Not going so far as to be able to support this hypothesis, he returns to the explanation of them being useful weapons in the struggle for life.

With the emergence of the interest in social and, especially, pro-social behaviors in the 1970s, new explanations, like those of Lorenz, supplant the old ones: the antlers are visual symbols of domination that will, through rituals, impress rival males and avoid direct combat. This solution, however, leaves the paradox of extravagance shrouded in mystery: If physical contact is avoided, what reason would there be for an advantage in large antlers? And what reason would there be for an animal to let itself be impressed by them? And why would females also let themselves be impressed by the imposing antlers, at least supposing, as Darwin did, a certain attraction for beauty? How, then, can the same ornament be a sexually attractive characteristic, and thus a burden—and go against the grain of natural selection—and constitute at the same time an effective structure in the frame of this same natural selection, as a weapon against predators and rivals?

NEW THEORETICAL FRAMES

These questions will benefit from the new orientations of interest for ethologists and new theoretical frames for pursuing them: for a vision of adaptation in terms of benefits (coming from the utilitarian tradition for which hyper-adaptationism—or panglossianism according to its detractors—is the successor) is substituted instead hypotheses in terms of cost.

This substitution of terminology did not for its part unite the paradigms of these theories. They are situated along two great theoretical axes according to whether they are adaptationist or, on the contrary, entertain the action of contingent and chance phenomena, lateral effects or mutations that are neither beneficial nor harmful. The adaptationist theories—including Zahavi's theory and the sociobiological theory—themselves give rise to two divergent historical traditions in the explanation of evolution and social behaviors: on the one hand, a pan-selectionism that I would qualify as static (like Zahavi's), and on the other hand the model called the "arms race."

For the supporters of the "arms race" model of evolution, animals are ceaselessly confronted with new problems and must respond to them by new strategies. There is always a lag between the problem and the response. This vision is clearly exemplified in Hamilton's theory of parasitism, the parasitized animal always being, as Richard Dawkins puts it, "behind in a war" on the parasite: to each defensive strategy of the parasite rapidly corresponds a new counter-offensive strategy of the parasitized animal. This theory can apply as well to the malady of myxomatosis as to the manipulations of the cuckoo. Another example is found in Krebs and Dawkins with the theories of "manipulators" and "mind readers," and in the most varied theories about lying in the animal world. The dupes come to develop a strategy that will allow them to detect the lie or the decoy (they become "mind readers"), that will in turn lead the "liars" to refine their own strategies of manipulation. This will drive the "dupes" into a situation where they are always "pressed into battle," so to speak.

This dynamic vision of the "arms race" makes us both witnesses and actors in the history of evolution: natural history is no longer the fictional history of a bygone past at the end of which everything was already completed and of which we would be only the passive heirs, but is transformed into a dynamic history in the full course of its unfolding. If this dynamic vision expresses in an implicit and optimistic manner the belief in constant progress, it is paradoxically founded on a sinister and pessimistic vision of natural selection. This is described as an incessant battle between rivals, in terms of strategies aimed at exploiting the other by any means possible.

It is necessary perhaps to pay attention for a moment to this renaissance of a Hobbesian vision of the world: it gives rise to a reciprocal renewed interest of biological and economic models for one another, "like the last time" (which is to say like the era when the Hobbesian worldview was held nearly unanimously by Western Darwinians, as it turns out the same era as Spencer). It could be the case that this is not a coincidence and that the economic models have a certain influence in those domains that borrow their tools and concepts. Economic models have a general tendency, in periods of crisis, to generate an ideal liberal discourse emphasizing the merits of competition, which would in effect be a return to the biological models with which they are intimately linked. This analysis is inscribed in the line opened up by certain French epistemologists—among them Patrick Tort, a student of Bachelard, Canguilhem, and Foucault—who dedicated themselves, in the 1980s, to showing that these types of sociopolitical influences on Darwinism could generate particular theoretical frames. These authors thus highlight what Michaël Löwy calls *elective affinities* that are created between Spencer's social Darwinism and the triumphant liberalism of England at the end of the nineteenth century that today still marks our manner of reading Darwin. But rather than this sociopolitical reading—although it is fascinating, and when the occasion presents itself I don't miss the opportunity to make reference to it—it is preferable, in the context of the theories that are contemporary to us, to consider a reading in terms of the affinities of the theory with our beliefs. Like the student who, with the nose to the desk sees neither the sentences nor the errors, the analyses of the stakes of a contemporary theory frequently require that, in order to be clearly understood, an intermediate temporal space be created that allows us to speak of these issues as having occurred in the past.

A HOBBESIAN YET MORAL WORLD

Recall that one of the most colorful beliefs that orients our fictional repertoire of natural history seems to be the belief in the moral order of the world. The cruel world of the nineteenth-century Darwinians eventually comes down to a moral world since an eminently moral law reigns there: the law of merit and justice. In its natural fictions, the battle for survival—amoral in its specific forms—thus obeys a moral law that presides over its general unfolding. Thanks to it, a purifying process is put in place: life will always go further and higher thanks to a ruthless battle and the elimination of the losers, extinction being the most complete expression of a lack of adaptation. When nature is no longer wise, justice intervenes, as in the world of Anaximander: the dinosaurs become too big, the elk antlers become too massive. Thus is illustrated the moral of a story that everyone knows: "His reach exceeded his grasp"

As in the myths, many levels coexist in these great moral fictions of the adaptationists, and this is the case from the level of affective representations of the belief in the justice of the world up to the representations coming from Christian visions of nature for which evil is present only to assure a better good. At another level, beliefs are supported by the moral of our epistemologies: there are "good causes" just as there are "good reasons." This is an important distinction in medicine, since it underlies, as in the remark by Isabelle Stengers,[49] the experimental starting point that separates the true causes from anecdotal appearances and the medicine of the charlatan. This distinction is frequently found in the sciences that tell stories: there must be good reasons in nature, not too much risk or fortuitous causes that would not match their effects. This idea of a moral that guides the course of natural history is found in the little incidents that pepper the theories—and play a non-negligible role in the context of justification: the term "selection" was rendered by the French translator of Darwin as "election" [*élection*]. The concept of election, which will not go unnoticed, must have introduced a resonance with liberal sociopolitical ideology, with Protestantism, and with the ideas carried by natural theodicies.

Election is also a moral cause at two levels: to begin with, it supposes the idea of an immanent justice (and we can think that this idea must have been consolatory for the narcissistic wound that had just been inflicted on humans); then, it reaffirms the ties between cause and effect at the epistemological level in amplifying the causal law: a cause is even more the cause of an effect when this cause doubles as an intentionality (which the term election recalls, as it is indeed the idea of choice).

The explanatory theories of the world will in this way most often use a particular rationality and obey certain laws of the genre. One of these laws is the presence of many levels of staging and symbolization in fiction. Each of these levels tells the same story, but each one at a different stage of abstraction: one level reveals the fiction, the story that is being told, the saga; another level describes our theory of knowledge; another relates the moral; and yet another makes manifest the ontology that founds our vision of nature and its inhabitants. Explanatory and scientific theories of the world revisit the mythic rationality described by Michel Tournier in *The Wind Spirit*,[50] and thus build stories in structures where each one of the stories revisits the same tale with a different system of symbolization. The particularity of natural history, such as we have seen it told here, is effectively the effacing of the distinction between two different levels that seem to lend it its force: the epistemological and moral levels appear, in effect, to become progressively indissociable since our epistemological beliefs seem to follow the rules set down by our moral beliefs. Causal laws and moral laws not only tell the same story, but the first appear, in the last resort, to obey the second.

LIMITS TO THE CONTEXTS OF JUSTIFICATION

Given all that has just been written, can we deduce that the affinities between a moral and a theory, or between a sociopolitical context and a theory, could thus explain the second term by the first, in exhausting the stakes, the sense, or the meanings? I propose that we consider the limits of this analysis and that we proceed by demonstration. The example of Kropotkin is in this respect an excellent illustration, since his theory mixes moral thought, political thought, and naturalist thought in an inextricable fashion.

Kropotkin's[51] zoological theory is generally seen as a more or less direct consequence of his political and moral philosophy, and of the utopia that he intended to build and legitimate.[52] Kropotkin refutes the Malthusian theories and tries

to show that animals use other strategies besides vital competition to regulate the problems of the division of resources.
Kropotkin thus collected an impressive array of observations
arguing in favor of the existence and prevalence of solidarity
behaviors in nature. Animals, Kropotkin tells us, have at least
two good reasons not to obey the law of the battle for existence
(in its narrow sense): First off, they are seldom confronted
with the utter lack of resources, and when they are they put in
place strategies of migration, which are themselves products
of evolution. The second reason, he tells us, is derived from
the observation that groups that preferred mutual aid to competition have better chances of survival and are thus favored by
selection. Kropotkin presents his arguments with a surprising
blend of anthropomorphic and technical arguments. The first
observations make us smile: Kropotkin claims that there is no
competition between the rabbit and the hare—this argument
of competition is elsewhere too frequently wrongly invoked—
but rather, the conflicts between them are due only to questions of character. As he observes, "The passionate, eminently
individualist hare cannot make friends with that placid, quiet,
and submissive creature, the rabbit."[53] He seems to prefigure,
some sixty years earlier, the arguments of Wynne-Edwards: for
example, when a group of beavers becomes too large to prevent the risks tied to overpopulation, the group separates in
two and one part heads up the river while the other one heads
down.

The authors analyzing this theory weave together the ties
between the political context and its scientific discourse and
show the extent to which Kropotkin only observed in nature
those phenomena that corroborated his hypothesis and his
utopia. The elective affinities that were created between Spencer and liberalism had imposed the vision of a Malthusian
and competitive nature; those established between the Russian Kropotkin and the anarchist ideal affected the collection
of data in the field and all but effaced the existence of conflict
around territory and resources.

In this same vein we could identify a good number of such phenomena in the history of ethology: Charles Carpenter described howler monkeys as a pacific and egalitarian community; thirty years later, after new observations, he described them as competitive and strongly hierarchical. Two ethologists describing the zigzagging path of two ants holding a prey between them will describe the scenario in two distinct ways: one ethologist will describe the ants as cooperatively carrying a prey that they want to share, whereas the other ethologist will see each ant attempting to pull the prey away from the other.

The context of justification would allow for the possibility of understanding both of these moments of interpretation, or at least would lead to a constructivist perspective of their analysis. But, in this analysis, one very important actor is missing, an actor who is obscured along the way since the critique generally dispenses with it: this actor is the very field or terrain itself, the physical matter, the events.

If it is true that they only exist, as the constructivists are quick to highlight, at the level of representations, the events cannot nevertheless have a representation without a field represented. Through separating the functions—the scientists in the field on the one hand, and the critiques of the representations, on the other—we end up at a rupture between those who speak and those who are spoken of, between the word and what we speak of.[54]

Daniel Todes made the following decisive remark about Kropotkin: don't forget that he is Russian.[55] This does not summarize a national, political, or social identity. His experience as a zoologist is drawn from his native land. This native land is a very particular one in relation to the one where Darwin drew his first observations: the Russian landscape is underpopulated with regard to its agricultural potentiality. The battle for resources should not take the same form there as it does in England. The difference of nature must have allowed Kropotkin to collect different data that did not in fact correspond to the Malthusian law, and observations of behaviors adapted to a different ecology. But this was not the only difference: Kropotkin traveled, and the difference between his native land and that of Darwin and Wallace—even if their observations were principally carried out in the tropics, the model of nature that seeped into them was an overpopulated world—made him attentive to the influence of the land on the theory that would like to describe everything.

Moral fictions that animate our ways of reconstructing the world are thus without doubt particular narrative styles, and it would be useful to consider them as such. But they are ones where the objects are themselves actors in their own right—and not simple figures, whose secondary role would otherwise allow us to tell the tale without them.

THE PROTAGONISTS OF THE ARMS RACE

Let us now return to the subject that we abandoned for this long digression, the arms race, to try to understand its purposes. To do this, we will situate ourselves outside the frame of reading in terms of the context of justification that tracks the affinities, such as, for example, those between our socioeconomic world and this ideology of permanent conflict as a source of progress. Rather, we will look at what the purposes of this arms race are. Astonishingly enough, its purposes, the ground where it emerges, are not so much the nest of the host of the cuckoo, the victim of the slave-maker, or even the body infested by a parasite.[56]

Behind their apparent identity—and sometimes surprisingly—is hidden a third term, the fictional objects of a particular laboratory: the simulation laboratory. And this laboratory has a more and more important role in the field of ethology.

The simulation laboratory created a new model of the economical animal and substituted, for an adaptive thought in terms of benefits, a model of optimization between a cost and its benefit, in insisting on the cost of each of these strategies considered. To this new level of animal behavior there correspond new tools: the mathematical tools of the game theory of evolution, whose existence we already mentioned regarding reciprocal altruism. Fiction constituting a mode of putting fictions to the test,[57] the mathematical modeling constitutes the putting into play of the hypotheses, the evaluation of their conditions of possibility. The initial economic model defined individual strategies linked by empirical relations of cooperation and competition, where the games are aimed at permitting each one to obtain, for its advantage, the largest differential distance possible from a given statistical regularity.[58]

The simulation of the fictional laboratory sets up fictions of relation: if A does this or B that, is the strategy of A (or of B) stable in these conditions? The model will essentially describe the ways to resolve conflicts (those of the arms race) and will gradually extend to all relations, even the most cooperative. Thus Robert Trivers defines the parent–child relationship in terms of differential investments: each one of the parents is in conflict with the other, with the goal of not letting herself be exploited by him, and even finds herself in conflict with her children, who want to draw all the possible advantages from the situation.[59] Each of the parents will thus adopt a strategy to assure the survival of their descendants and eventually—if it is in the practice of the species—the cooperation of their partner. At the same time, she tries to avoid overly costly investments for her own survival and reproductive success.

The objects of computer laboratories are virtual objects in the processes that create fictions but can also be seen to receive a real referent on the ground; nominal strategies themselves generally designate their referents in numerous cases: the parental strategy possesses a number of referents in the real world, even the strategy of the cuckoo.[60] We will have occasion to return to the role of these laboratories in putting theories to the test.

Another vision of evolution continues to oppose itself to the fictional and mathematical rationality of the simulation laboratories. This other vision takes the form of mythical rationality in a history that has become static because it is already revealed: each trait, behavior, or structure here becomes the product of a history that gives it meaning and makes of it a useful and adapted trait.

Each of these traits deserves and receives a particular reading, an interpretation that explains its utility. According to this grid of interpretation of the world, a trait or a behavior cannot therefore be the consequence either of a delay of adaptation or of an error of the animal: the response of one who seems "deceived" by the other (the host of the cuckoo, the dupe of the lie) conveys "the best compromise possible"[61] between the act of endeavoring to avoid error, at the price of an elevated cost (throwing all the eggs out of the nest if you suspect that a cuckoo has visited), and the act of falling into the trap, at the price of another cost.

In my opinion, these two theories—even if only one of them can hold its own—present two versions of optimism: the first is founded on the conception of a constant progress, and therefore advocates a future optimism, the second, founded on a more static equilibrium, posits that the best possible is already here.[62]

2. Rituals between Altruism and Reproductive Function

Of the two paradoxes mentioned, altruism and the extravagance of signals, neither was totally resolved. Partial answers were made, which nonetheless left inexplicable residues: why doesn't an animal cheat in a cooperative relationship? Sociobiology only accounts for final causes and doesn't cover the field of interactions between non-related animals. Why is an aid that sometimes appears as more harmful than beneficial to its recipient maintained?

How, in the frame of sexual selection, can the same characteristic end up taking on the double function of attracting females and intimidating rivals? Why doesn't extravagance tend to disappear?

It is this type of question that the theory of signals, and its corollary, the handicap principle, address.

Its originality, as we have already mentioned, consists precisely in the fact of bringing one single response to two series of questions: altruism and the signals selected by sexual selection are two modes of expression of the same process. This takes on its most specific meaning when we consider the question of the function of these rituals.

Effectively, the theory of the ritual is situated, for many reasons, in the frame of theories of altruism and sexual selection: as a communication behavior, ritual belongs both to the domain of cooperation and, at the same time, to that of partner choice. Ritual is a behavior that can favor cooperation in a double sense: it permits the restriction and inhibition of aggression, on the one hand, and underlies elaborate forms of cooperation, on the other hand. Among these—and in this respect it belongs to the domain of partner selection—we find cooperation that represents exchange in anticipation of procreation. This cooperation is itself also facilitated or even conditioned by the enactment of rituals.

Julian Huxley[1] defined the process of ritualization as the evolution of movements that had lost their original function to serve as signals. Through his research, Lorenz was able to reveal that ritualized manifestations are more stereotyped than the original movements from which they are derived, this evolution being the result of a common interest of the communicating parties to do so in the clearest manner possible. If the stereotype augments the clarity, it however diminishes, according to Desmond Morris,[2] the information carried as to the state of the signaler. He then formed the hypothesis that the signal developed a certain variance in its duration, intensity, and repetition in order to compensate for the losses of information.

In depicting this intuition, the lion's share of observations accentuate what, in the ritual, allows clear comprehension of the intentions of the parties without ambiguity, which is to say what constitutes resemblance and stereotyping at the heart of languages and rituals.

RITUALS AMONG PRIMATES

In recent years, research in primatology has found itself confronted with new questions that its subjects of study imposed on it. This led primatologists to become interested in the particular forms that certain rituals can take. They also tried to understand how a specific message can be tied to a particular gestural form.

Male baboons *Papio cynocephalus anubis* present gestural exchanges including highly ritualized sexual acts, such as the ritual presentation of the posterior, the presentation of the penis, ritual mounting, or touching of the genital organs. These interactions contrast with the habitual behavior of male baboons, among whom competition seems to be fairly crude and the source of numerous tensions. Despite this, a baboon can greet another one in showing the most vulnerable parts of his body, and in permitting him to touch his genital parts, thereby placing, literally, his chances for future reproduction in the hands of a potential rival. To explain why these gestures—or more rigorously stated, to explain how these rather dangerous gestures are maintained when more "anodyne" gestures serve the same purpose among many primates— Barbara Smuts and John Watanabe tried to evaluate the particular contexts in which these rituals took place, as well as the variations that they showed.

The rituals generally took place in a neutral context, which is to say without a particular tension and without the possibility of conflicts around a resource. The baboons had woven quasi-amicable relations of cooperation in protecting one another against the aggression of rival males and in mutually aiding one another to lead a female well away from the troop, showing the most complete and intimate rituals of the behavioral register.

Status is highly variable among baboons who are part of the same age group: the rituals probably have a function of "testing" the relations of dominance by less damaging means than direct conflict. But, between an older and a younger baboon, status is clear, and does not need to be tested or reinforced. It is thus possible that they use ritual instead to explore the possibilities of developing cooperative interactions.

Baboons can mobilize the possibilities of ritual in this way to negotiate and build some aspects of relation, including cooperation. In the final analysis, every real cooperative relation depends on the mutual accord of each partner. The one who engages can never be sure that the partner will not withdraw. Given this possibility of defection, vulnerability in the face of a possible betrayal will always temper confidence, and each individual really desiring to cooperate will be confronted with the problem of knowing how to convince others of his good intentions. Before each baboon can benefit, in quietude, from cooperation, it is necessary that this mutual cooperation be well established, and that each party be convinced that the other will resist the opportunistic temptation for immediate benefit. Baboon rituals perhaps serve this function by formally establishing, through gestures and mimicry, a neutral and clearly structured context in which the animal acts in an invariable manner by following rather than creating the sequences of their performance.

This quasi-theatrical conformity is a mutual conformity shown by each of the partners that would constitute, according to Roy Rappaport (cited by Barbara Smuts and John Watanabe), a paradigm for social cooperation, since it "so simply, yet immediately and unambiguously, subordinates individuals to common behavioral constraints that transcend the individual actors themselves."[3] In this way, the minimal social coordination necessary for the ritual can feed the possibility of future mutual interactions, in the same way that the refusal to participate in it will be inevitably clear to everyone.

The intimacy and danger tied to this type of closeness of behavior are thus not without meaning. Lorenz, for example, interprets presentations and mounting as gestures of submission by the one who presents, and a recognition of this submission on the part of the one who mounts, an acknowledgment that can for that matter pacify the partner who is defeated in a conflict. For Smuts and Watanabe, the reason for why baboons place their future reproductive chances in the hands of another male through a pact has more to do with the fact that baboons do not have at their disposal more complex communication and are incapable of taking the oath so they instead have recourse to a gestural equivalent of this swearing of the oath. Such a risky gesture can help to build this confidence to the degree that it imposes a highly elevated potential cost.

For Smuts and Watanabe, taking the oath among humans serves the same end—that is, to accept submitting to punishment in the case of a breach. According to them, certain anthropologists described similar rituals among the Aboriginal Australians who show their promise to help someone in presenting their penis to him. To confirm their interpretation, the authors appeal to the common origin of terms designating the fragile parties presented, the testing of confidence and the oath: the *Oxford English Dictionary* suggests, indeed, that the words testament, test, and testicle come from the Latin *testis*: witness. This would also give *testari*: to witness.

The examples collected among the Aboriginal Australians seem to me to arrive a bit hastily to the rescue of a hypothesis postulating a sort of "semanticization" of the gestural. Beyond that, even if the Aboriginal Australian presents his penis, this does not signify that this custom would be universal, nor that the term designating the promise, in his language, would have an etymological rapport with the term by which he refers to his organs. In parallel, the proximity of the terms in English and French does not permit us to affirm the existence of a similar custom in the past.[4]

The attempt to find a theory that re-creates the tie between the container and the contained, between the form and the content, between what we express and the manner in which we express it, is nonetheless very seductive. There is an important issue at stake here. It concerns something other than simply finding the "unity of the real": it becomes a question of re-creating the link between the human and the animal, of discovering the precursors of language. This interpretation allows for the resolution of the problem concerning a line of continuity between humans and animals, since in showing that language is not totally arbitrary, we can imagine finding its origin in the gestural precursors among primates.

Despite the sympathy that we can feel toward this type of project, an interpretation determined by the semantic analysis seems to me, in this area, inadequate. However, the commentaries of the authors bring an interesting clarification to the degree that they show the specificity of the ritual gestures, their singularity, and what would seem to be the necessity of executing them in a dangerous manner. This opens the way to an alternative hypothesis to the classic idea according to which aggression is present in the ritual as a "residue" of the ritualized compromise. This hypothesis of Lorenz holds that aggression is implicated in the relations of social animals, first of all because they are adaptive, and then because individuals cannot be selective with their aggression on first contact. Subsequently, the function and meaning of numerous rituals have been interpreted with this explanation in mind: they inhibit the aggression of partners and allow the rapprochement necessary to the establishment of ties. Consequently, and despite their function of inhibiting aggression, they will, in the formation of these ties, always in part reflect this aggression, either by directing it toward another member of the group, or in the member's absence, toward an imaginary enemy, or by simply sketching an outline of the inhibited aggression in a rather ostentatious manner without ever bringing it to its full completion. This is what is called, in the context, the residue of compromise. In this way, swans greet their partner by lowering the head onto the back and clicking the beak. This is, according to Irenäus Eibl-Eibesfeldt, a signifying demonstration of the "rerouting of aggression."[5] The attack of the enemy, even imaginary, and the ensuing cry of victory among the greylag goose reflects, for Lorenz,[6] this process of ritualized reorientation of aggression.

Between an interpretation of ritual in terms of fixed automatisms and an interpretation in terms of quasi-linguistic communications, a middle way that takes account of certain of its characteristics can be traced.

RITUAL AND THEORY OF SIGNALS

Already in his very first articles,[7] Zahavi reconsiders the function of aggression in the rituals for the formation of ties, and grants them a larger function than catharsis and the communication of appeasement. If we define the tie as an accord in anticipation of a coalition, the formation of the tie must, according to him, bring information to the one who carries out the rituals.

This information must inform the partners about the trustworthiness of the other, his motivation, and his availability. Now, a relatively rapid means of evaluation of the availability and trustworthiness of a congener can be tested: an individual truly motivated to invest in a relation will accept a stressful test, a more reluctant individual will refuse it. In this way, the ritual greeting of the baboon is explained, from this point of view, as the act of "testing the tie." This hypothesis not only takes account of the particular form of the ritual, but also offers what seems to be an enviable quality of explanation for the ethologists: one single hypothesis takes account of both the form of the ritual and its meaning.

This hypothesis of a "test" function of the quality of the tie in the midst of stressful situations leads Zahavi to reinterpret the numerous observations of behaviors until then judged as accidental or out of control: a territorial male bird, in threatening a female who pays him a visit, sometimes thwarts the possibility of a coupling.[8] According to Zahavi, the male bird in fact assures himself, through undertaking this stressful test, that the female does not come only to take advantage of the visit in order to feed herself. The aggression shown toward a new member of the group emerges from this type of exam.

In this way, the danger inherent in the gestures in some of the baboon rituals supports the hypothesis that the crucial function of ritual would be to put confidence to the test. If we extend Zahavi's reasoning to other situations, we could consider that, beyond the reinforcement of ties due to the exchange of goods, one of the components in social grooming is the evaluation of the reliability of the other: one of the two partners, in effect, turns their back and totally relaxes. He is thus in a situation of pronounced vulnerability.

Creating ties or reinforcing them involves applying a series of ritualized signals signifying the desire to establish a relation. In this case, as numerous ethologists have already remarked, the "risky" gestures inciting confidence will be privileged, like the act of offering open palms, lowering the head in presenting the vulnerable part of the skull, offering the genital parts (a gesture that is also present in the rituals of other species, as among the spotted hyenas), opening the eyes wide and enlarging the pupils in limiting clear vision, suppressing this vision by the act of crying when one implores the other, et cetera.[9]

The fact of, on the one hand, accepting the risk, and on the other hand not taking advantage of it, thus constitutes, in Zahavi's eyes, a function that affirms and increases the confidence between the two partners.

In parallel with this function of testing the credibility of motivations or intentions of a potential partner in relation, the ritual fulfills other functions: it permits the evaluation of a whole series of additional essential information like, for example, the qualities and the particular motivation of a potential adversary. To measure the originality of this theory, we must keep in mind the fact that Lorenz and Morris insisted on the stereotyped, specific, coded—we could say conventional—aspect of gestures performed in immutable sequences.

Now, the loss of information of what would constitute the singularity of the participants in favor of the clarity of a common language poses some problems if we try to understand the outcome of some instances of ritualized combat.

If we reconsider the example such as it was formalized by the game theory of evolution, a question persists: how can combat result in the victory of one of the two protagonists if, thanks to ritualization, they never come to touch one another? The "hawk–dove" model tries to respond to this question.

To begin with the model stipulates two strategies of combat: the strategy called "hawk," which is a ferocious non-ritualized strategy, and the strategy called "dove." The "dove" organisms of this hypothetical population only threaten in a conventional manner, never attacking or injuring. If a "dove" encounters an organism that adopts the same strategy as it does, no one is hurt; they continue to intimidate each other until one of them has had enough, or decides to no longer bother with the process and flies off.[10] The only decisive factor of this type of combat will in the end be that of "time." An animal must feed itself, and thus cannot surpass certain limits of time in ritualized combat. But if this is the case, why are there such parades of intimidation since, in the final analysis, only time is decisive? The question comes back to asking how an animal convinces its adversary that, if it suddenly changes strategy and becomes a "hawk," the ensuing combat will be fatal.

To respond to this question, it is necessary to see the ritual as *a set of gestures that express differences.* And to do this, one must consider the stereotyping that the ritual represents in a different manner, no longer considering it as information in itself, but rather as the frame for information.

THE RITUAL, THE MARTIAN, AND THE MARATHON

Seeing differences where similarities are usually observed is what, in my opinion, situates and establishes the originality of Zahavi's approach. We can create an example based on one of his metaphors that will help us in our progression through the science with a bit of science fiction, but it will not take us too far from our subject.

Imagine that an ethologist, similar to our ethologists in terms of theories, ways of thinking, and field practices, arrives from another planet to observe the inhabitants of Earth. She lands in the center of the country right in the middle of a national highway, which, as it seems to her, must have been drawn by the passage of numerous organisms through this enormous human habitat. She decides to apply herself to studying them, and thus to follow their tracks. At a surprisingly regular distance, on the side of the road, panels appear on which strange signs are inscribed, always the same: "km" [kilometers]. Our ethologist concludes from this that the inhabitants of this human habitat mark the paths they follow with a remarkable frequency, in order to remind the organisms of their species that they have themselves passed here, or that they follow a path of the grand passage. Our ethologist comes to the crossing of two dark gray paths and encounters a group of humans. This group is running a marathon, but our ethologist has no way of knowing what a marathon is since it doesn't exist on her planet. We could either think that there is no competition on her planet or opt for the opposite hypothesis: there is so much competition on her planet that our sidereal ethologist is intimately convinced that it cannot exist in what represents, for her, the otherwordly. But here we make digressions that don't have too much to do with ritual. The ethologist, in view of all these organisms performing exactly the same movements, in an almost perfect cadence—she learned, in an anthropology course, that error is human—with such crowding and unanimity, could, in a very plausible manner, be convinced that she is observing a ritual for the calming of tensions or the reinforcement of cohesive ties among the group. And she would be partially correct.

But she would miss the function that the marathon serves for each of the organisms: she separates the differential qualities of the runners in seeing them as a standard. In effect, if everyone does the same thing, this is to allow for the truly significant differences between the runners to emerge in a clear and simple way.

This can also be said, according to Zahavi this time, with regard to ballet competitions: the gestures of the competition are standardized, so that the jury can decide the qualities of each of the dancers participating in it.

Saying "good morning" relates to the same functional utilization of the standard. "Good morning" does not give any information—and I don't know anyone who bears in mind, each time they say it, the real signification of the words—but the way in which we say it when we greet others constitutes information that is really conveyed: it indicates to others our mood, our involvement in the relationship, our desire or pleasure in seeing them, and even our morning energy.

From this perspective, the ritual acquires a new function: it not only puts in play, as in the classical theory, the mechanisms of recognition of an innate language, but above all the competence of each individual to perceive, beyond the similarities of the ritual, differences in motivations, the content of intentions, trustworthiness, and even the qualities that identify each of the participants.

To this competence of individuals participating in the ritual corresponds, like a mirror, that of the ethological observer. The new definition of the ritual takes the allure and the arguments of the classic controversies in ethology, that is to say that the focus of interest on stereotypy was in fact nothing more than the consequence of an inadequate observation: what we saw was not what should have been seen. The controversies arising in ethology frequently use arguments questioning the competence of the observer to create the proper conditions of observation. The analysis of ritual, even before the controversy can be declared, will get ahead of the arguments. The analysis will claim, in an explicit manner, the fact of being founded on a more pertinent and detailed observation in order to refute the classical theories.

This new definition of ritual that establishes an identity between the competences of the observer of the system and the competences of the observed (become observer in the system) creates what we could call a mirror effect between what we observe and what is observed, an identification—about which we cannot yet know if it is metaphorical or psychological—between the different actors of the system.

To this metaphorical or psychological identification—this mirror effect that is produced between the observed and the observer, and that we will find other occurrences of elsewhere—corresponds an effacing of the boundary between the content and the form of analysis of the material observed: each behavior becomes, from this perspective, an exchange of information. This is not new and comes up each time the systems observed are analyzed as systems of communication. There is a mirror effect because the same competences are required for the one and the other, because the observed becomes in turn the observer. There is a mirror effect because the observer and the observed are both pressed into the same work of decoding behavior into information: male aggression toward a female is no longer a simple aggressive behavior, but decodes as a question that he poses to her; social grooming is no longer simply what creates ties, but becomes a descriptive and predictive exchange that defines the relation; and so on. A behavior is no longer a simple action on the environment, but becomes a series of pieces of information that the individual gives to his environment. Each behavior is therefore capable of taking on a signal value. A distinction must however be made between the concept of information and that of signal: if all signal is information, all information does not constitute a signal. A signal is distinct from other characteristics that are bearers of information and from the emanations of the inanimate world: only a characteristic that has been shaped by evolution in the response that other individuals have made when they perceived it is a signal. A signal is, therefore, a characteristic that was selected thanks to the response of other individuals to this characteristic in the course of evolution.

EPISTEMOLOGY AND ONTOLOGY OF DIFFERENCE

A vision that encourages the perception of differences beyond similarities, that thinks information in the Batesonian terms of a "difference that makes difference," that considers similarity as a tool of differentiation and not as an effacement of differences in favor of a common denominator, gives a quasi-ontological priority to differences—a quasi-ontological priority to the degree that the fact of simply according an epistemological priority to difference comes back, definitively, to according to the "same" a status equal to that of the different, to begin with. It is a matter, as in the vision of Necker's cube, of two ways of considering the ritual, of which one seems, for one or another reason (the reasons for according an epistemological priority), better than the other. The quasi-ontological priority of difference is linked only to it and relegates similarity to the status of a tool—the real *is* difference, similitude is a mode of creation of differences. This way of thinking can also be related to a shift that operates between a classical tradition, seeing the ritual as the moment of cooperation par excellence—the moment of cohesion, of coordination, of pacification, of the inhibition of aggression, symbolized by the harmonization of the figures—and a new, while perhaps also remaining very ancient, tradition that considers every relation in terms of competition or conflicts of interest—in sum, of *differences.*

The ritual is a test guaranteeing the reliability of the claims of the partners in the relation: claims to superiority in the case of ritual combat and a test of the reliability of intentions in the preparatory rituals for cooperative relations. The metaphor of the marathon is less than innocent here, but this shift of conceptions becomes most evident in the example of ballet competitions: what we describe as a cooperative enterprise executed with the goal of creating beauty in harmony becomes the standard that allows deciding between the competitors.

Such is the case with the dance of the Arabian babbler, at least for some of its spectators.

PART II

The Dance of the Babbler

3. The Arabian Babbler

We have spoken about the dance of the babbler in foreshadowing that it would be the knot of our story and that we would follow the indications of the threads untangled in our labyrinth. Indeed our path will now take on the contours of a labyrinth. Better than a thread or a web, the metaphor of the maze will represent the spatial and affective configuration of the aporias in which we will engage ourselves, and that we must cast aside along the path of our journey.

I had already seen birds dance: dancing is also the name we give to the rutting behavior of the great crested grebe that, during the formation of a couple, will perform gracious movements in front of its partner. But I could not imagine, from the simple reading of Zahavi's articles, that I would witness such a surprising and entertaining event: the group dance of the babblers. The babblers form a line of several individuals, a bit in the style of Russian dancers, and will press in toward one another until the point when, suddenly, one of the birds leaps over another or several others, and inserts itself between two companions, more toward the center of the line of dancers. They change places in this way one after another, by inserting themselves as close as possible to the middle of the line. Sometimes, the line can transform into a kind of closed ball, the middle of which each one again tries to occupy as the central position. Each bird joins in, in an organized crush that mixes all of them, alternatively in different places on the outer edges or in the center of the line or circle that is formed. The dance can last more than thirty minutes and involve the whole group. It generally takes place in the morning, at crack of dawn, when everything is still in semidarkness. Occasionally, the birds can dance at sunset. And every now and then another dance ritual brings them together, when conditions allow a communal bath (after there has been adequate rainfall to form puddles on the ground): just after the bath, the babblers begin to dance.

What is the meaning—that is, the function, since in the field of natural history the meaning of a behavior is nothing but its function, "to mean" becomes "to be good for"[1]–of the dance ritual?

CONFLICTUAL INTERPRETATIONS

Is the dance ritual, as Zahavi suggests, a test imposed on the partners? The shoves and crushes, the ejection of those who are at the outer edges, the taking of another's place, are, from this point of view, the stress imposed on congeners that permits each of the participants in the ritual to evaluate how much the others are motivated to cooperate. Not only does it appear to be a way of testing the relation to a specific other in a stressful situation, but also of evaluating the strength of the tie that binds each of the members to the others. When a babbler finds itself "held" in the dance—both literally and figuratively, the bird is really "stuck" between two partners, and more particularly still when it occupies a "coveted" place—it strongly restricts its possibilities to take flight. What's more, the time of the morning dance is the worst possible moment: the sun has barely risen, the babblers have great difficulty seeing in this penumbral light, and they represent easy prey for a predator. Furthermore, after a long night of sleep, the birds are hungry and in need of seeking out a source of energy. Even the site of the ritual is just as surprising: the babblers usually dance in an uncovered area, where they are the most vulnerable to a potential predator.

According to Zahavi, the place and the time of the dance are the clues: they are chosen precisely because these places and times present the greatest difficulties.[2] It is this difficulty itself that gives the meaning, the function of the dance: it allows verifying the reliability of the motivations of the group members. The babblers of a group who "are willing to sustain a cost imposed on them by other group members do so because they have a desire to be group members."[3]

The theory of Zahavi's assistant Roni Osztreiher responds to this hypothesis and subjects it to serious questioning. According to Osztreiher, the dance of the babbler must be seen as the expression of a competition for demonstrating one's superiority over the others who are dancing, with each secured place in the dance considered a victory, and where the quality of the bird who is capable of securing and holding onto the central position the longest is evident. The clearest indication for this function of the dance is that it generally takes place at a time when the division of the resources of the group is decided: females during the reproduction season, food resources before the winter, et cetera.

The dance that, for Zahavi, constitutes the preliminary groundwork for cooperation becomes, for Osztreiher, a mode of competition. In this latter case, as we will soon explore, the dance would become the performative expression of status.

The babblers sometimes play a game that reproduces the same type of scenario, but in a space not delimited by the line, and not ritualized. Each bird occupies a position (on a branch or on a rock), and then changes positions in order to occupy that of another bird by ousting them from it. The game appeared to me as a variation of the dance, or perhaps as its preparatory exercise. In a reading that hews closer to the definition of ritual, we would have to observe that the dance constitutes a rerouting of the gestures of the game for the benefit of communicative functions. However, neither one of these alternatives is fully satisfying: the reason for this no doubt resides in the specificity of the two behaviors. In a relatively straightforward manner, both alternatives fill up the behaviors with metacommunicative functions: they "speak" on the relation and define it—whatever the interpretive hypothesis envisaged, whether stress or the expression of superiority.

It would seem to be hazardous to opt for one option, claiming that ritual borrows its form from the game—in the same way that the ritual formally adopts alimentary or aggressive behaviors to reroute them from their initial functions—or for the other, claiming that the game derives from the ritual. Whatever it may be, the game itself could support the two interpretations that contest each other for the explanation of the ritual: like ritual, it can constitute a test of the reliability of the partners and of their capacity to withstand stress; and as site of competition, it can be the occasion to recall or test the strength of each one. Regarding the ritual, as regarding the game, each of the interpretations, with observation, seems equally plausible with regard to the facts.

Notwithstanding these questions, this controversy was never discussed in an explicit manner, as if it was already clear, in a certain way, that the interpretation was a matter of choice. Zahavi did not explain to me the reasons for his position with arguments that could have separated the two propositions. His assistant had not published his own arguments, and so I was only aware of Osztreiher's hypothesis. He gave me two different explanations at different times, in very different styles: he provided me with the first explanation when I asked him about his hypothesis while we were working together; the second explanation came several months after my return from fieldwork, following a long letter in which I had asked him for other clarifications concerning the editing of my book. The response this time took the more conventional and developed form of an extract from his article.

When I asked him to better justify his representation of the dance, he gave me a response that I didn't understand until much later, in light of other elements that I had collected in the weeks that followed. We were seated next to one another in front of a group of birds, when I asked him to set out the two interpretations of the dance for me: he began by raising and placing his arm over my shoulders, and asked me, "How do you interpret this? As a stressful gesture?"—and, in certain way, I experienced this gesture, on the part of a stranger, as stressful—"Or as a gesture by which I demonstrate my superiority?" "And now," he said in pushing me somewhat violently but without malice, "how do you interpret my gesture? Everything is there. That's how the two interpretations are."

I tried to watch the dances with each of the two hypotheses in mind. In turn, both seemed equally plausible, and the facts fit both appropriately. I was able to uncover clues and information that would conform to both hypotheses. A pragmatist in the style of William James confronted with the problem of the squirrel would say that they were equal and that nothing, in the facts, would favor choosing one or the other.[4] He would add, no doubt to close off the debate, that it is not the facts that are at play here, but the links that we make between them, and the categories that we sketch out to better understand them.

What could have cried out to me was however rather simple: Osztreiher's example was not thrown together haphazardly, and apparent contradictions (the caress and the shove) did not have the function of opposing the two hypotheses in a symmetrical manner. Each of his gestures was aimed, deliberately, at supporting the two interpretations with the same degree of conviction, in an equitable manner. What I couldn't make sense of at the time, without the perspective of elements from later analysis, was that his example did not speak to me about the babblers, but about our singular relation when confronted with them. It seems to me now that it was more important for him to defend a language community that we could share rather than to convince me of the validity of his interpretation of the dances. It is necessary to note that Osztreiher's research primarily concerns the observer effects on the dances of the birds. We will return to this at greater length later. The "influence of the observer" refers to, on this account, the statement [énoncé] of a constructivist hypothesis. But this constructivist hypothesis can be qualified as a "soft" constructivism, since it considers the effect of a relation on an object that we do not doubt offers a stable and real support to the observation. The real is given to us, but we modify it. In other words, the bird is what it is, but its behaviors can be modified by the fact of my presence. I cannot therefore know how it is independent of me, independent of the knowledge that I have of it. To this "soft" constructivism one could oppose a more radical constructivism—and one closer to my initial ambitions—that addresses how the construction of the object itself by the active observer subject takes place. Here the real is hardly presented to us or perhaps not at all. We merely construct a representation of it. I had defined Osztreiher, in a somewhat summary manner and by virtue of the experiment that he conducted, as a "soft" constructivist, preoccupied by lived experience and inquiring into the effects of his presence near the birds. Today I think, in light of certain events like Roni's passionate interest for all that touches on the anthropological aspects of my research, that he is, in fact, a much more radical constructivist than I had imagined at the time, an enthusiastic constructivist when he evokes the idea that a particular gaze can yield a particular representation.

When I asked him to give me his interpretation of the dance, Osztreiher responded to my question, but in a way that I didn't expect. I hoped that he would give me some content to analyze, that he would provide me with the elements that would allow me to unravel the links between his methodology, his personal myths, his gaze, and his hypotheses. He presented me with something else altogether: he offered me what I had already defined as a community of language,[5] he offered me a hypothesis, he showed me that he knew what I was looking for, and that he was looking for it with me, in implicating me directly, physically—the intrusion into my space, the transgression of individual distance is not, in my opinion, without meaning. Above all, he showed me that he refused the role that I had submitted him to: that of the observed. Osztreiher did not therefore respond to my question as the observed would to the observer that analyzed him; instead, he suggested that I actively construct a frame within which we could inscribe a dialogue regarding gazes on the babblers. In asking me to interpret a gesture, Osztreiher thus put me in front of the two branches of a choice: be myself the observed who gives the interpretative content to his gestures, or build with him a common frame of reflection in which we would both be observers. The interpretation of the dance was nothing, from this perspective, but the pretext for the modification of our relationship.

If it is true that neither Zahavi's nor Osztreiher's interpretation of the dance has any true effect on what is taking place, what does change, in their particular relationship, is one of the most important elements of the context of justification for the theory: each one, the assistant and the professor—for reasons that are difficult to understand—chose to stick with his own interpretation. The article that was initially meant to be the fruit of a collaboration—and benefit from the support of Zahavi in scientific journals—would end up having only the assistant's name on it and traversing a more difficult road to being published.[6]

This reading in terms of conflict of interests putting the cooperation of the two scientific partners at stake—where the status of one can really influence the status of the other—returns us to the conflict of interest that, among the babblers, makes rituals so necessary. Whichever of the two interpretations one chooses to rely on, the ritual emanates from a conflict of interests: otherwise, what need would there be to evaluate the motivation or the reliability of the partners, on the one hand, or to demonstrate one's superiority, on the other hand, in a relationship where no conflict is possible?

By way of our preceding explanation, we are already fully in the Zahavian optic: the initial postulate of this student of David Lack, fervent champion of individual selection, is simple enough. Every cooperative system, whatever it may be, presents a dilemma to its members. Each of the individuals is, in effect, confronted with two contradictory pressures: enter into competition to assure its own better reproductive success, on the one hand, or collaborate, on the other hand. In watching the babblers build and manage their social relations, Zahavi would come to elaborate a surprising theory with somewhat of a hybrid aspect, since it seemed in certain moments to flirt with the theory of group selection, then to reject it and take up the hard law of individual selection again: "In a system in which individual survival depends on the health of others, there is much to gain for all, even the strongest and most dominant, in avoiding combat."[7] It is not so much a matter of assuring the life of the group for itself, but of doing so for the survival of the individual. The group is thus not an end, but a simple means.

Arabian babblers, notwithstanding what was just written, are extraordinarily altruistic and social birds. Not only do they dance and play, but, as we have already referred to, they regularly give each other presents, offer each other food, groom each other, participate in the raising of young as helpers at the nest, watch juveniles with careful attention even if they are not related to them, take on the role of sentinel with remarkable constancy, sound the alarm in case of danger, and can even attack a predator to save a congener in danger (mobbing reaction). Extraterritorial conflicts against other groups are extremely harsh and also constitute occasions for great bravery and heroic gestures. On the contrary, few conflicts seem to arise at the heart of the group, since the babblers, Zahavi observes, dedicate much time to altruistic behaviors.

How can it be, we might say to ourselves, that such cooperative birds did not succeed in converting a theorist of competition into a fervent defender of group selection? Naturally that is the whole question. Do we finally find ourselves here confronted with a good example of the work of ideology? Confronted with an effect comparable to that of the shattered mirror from the story "The Snow Queen" by Hans Christian Andersen, a mirror that accidentally slips out of the hands of the devil and transforms the gaze of all those who had received the shards in such a way that the beauty of the world no longer appears to them. Ideology would then be like the shard of the mirror that, in the eyes of Kay, "brings into relief the foul and nasty side of beings and things and accentuates their faults."

DETERMINATION OF THE GAZE BY ITS CONTEXT

We have already said that Zahavi's gaze can pay heed to singularities and differences beyond similarities. Can it, at the same time and without contradicting itself, obscure the object, silence it from speaking in its place?

If we revisit Latour's idea that objects, whatever they may be—the hole in the ozone layer, the HIV virus, frozen embryos, sensor robots, and even, why not, evolutionary games— are hybrid objects that refer simultaneously to nature and culture, we could then think of the "babbler in discourse" (Zahavi's discourse as it turns out here) as an object that is also hybrid, held in the threads of a web that we call the context of justification, composed of beliefs, of mentally represented social structures, made of emotions, of selected facts, of the links between these facts; made, too, of an Israeli society with its kibbutzim, its unending war and its kilometers of threatened border, with its egalitarian utopia and the sense of bitter aftertaste, with the feeling of emergency and daily loss. But this hybrid object that is "the bird in discourse," or "the bird of the Jordanian border," or yet again "the bird of a Palestinian pioneer," is also simply a bird, irreducible to a series of social facts—and who itself actively participates in the theoretical discourse that it is the object of.

The babbler participates within the discourse by offering researchers a certain particular manner of behaving. In some ways, it resists the traditional attempts at explanation. The "babbler in discourse" represents, in a certain respect, the happy conjunction between a series of anecdotal behaviors that apparently have nothing to do with one another and a particular gaze seeking other types of responses for the questions usually posed to the birds.

A series of observations of a heterogenous nature appear in the articles, starting in 1974. At that time, observing that helping at the nest is more harmful than helpful, Zahavi is led to reformulate—in terms of adaptation—the question of its maintenance. He ends up producing the hypothesis that these partners play a beneficial, even imperative, role in the defense of the territory, and that these benefits can explain the babblers' social behaviors.[8]

In 1976, he notes that the manner of taking the babbler in hand can provoke different and paradoxical reactions for its part: the bird responds to a light grasp with aggressive behavior, and to a brusque grasp with calming behavior.[9] These behaviors are interpreted by Zahavi as modes for the management of conflict: the light grasp must be understood by the babbler as a behavior of submission that prompts a domination response, the brusque grasp leads to submissive behavior on its part. Other observations show that the babbler will not react aggressively if a lower-ranking male interferes in the act of mating between a female and a dominant. We should specify that the babblers never mate in the presence of another member of the group. Consequently, any subordinate male who decides to prevent a mating has only one thing to do: show up at the fateful moment. Thus the stake here is an important one. According to Zahavi this requirement of privacy would be the act of the female. Thanks to this tactic a female can observe the capacity of the male to control the other members of the group in preventing them from interfering in the relationship. In other words, the future father of the clutch must show the female his aptitude at protecting the eggs against possible saboteurs. However, when a congener effectively interferes in the relation, the reaction of the dominant thus thwarted is surprising, even paradoxical: he resolves the conflict by much more subtle means than threatening or aggression. The alpha male approaches the beta male, offers him food, and will even go so far as grooming him. The only response the subordinate can expect is the "cry of reprimand" launched in his direction.

It would seem, Zahavi then writes, that these modes of management indicate that the babblers have established a compromise between the conflicts necessary to the establishment of hierarchy and their inhibition, itself necessary for co-operation within the group.

In 1977, the particular forms of altruistic behavior appear to him: not only are they not done by accident, but in addition they are oftentimes followed by strange responses.[10] When a congener presents him a gift while emitting the little whistle characteristic of the gift-giver, the happy beneficiary gets mad and attacks the poor altruist. With regard to another bird who offers to relieve him from his role as sentinel by flying up to a branch below his, this temperamental babbler will show the same aggression. Similar observations add up. The figures bring some precision here: 99 percent of behaviors of offering are made by the dominant toward a subordinate. The others are, like those we have related, followed by conflicts.[11]

To the contrary of what happens among many birds, for babblers it is the giver who emits a particular signal during the transmission of food. These gifts are usually made in open spaces, and everything leads one to believe that they are real exhibitions. I open a brief parenthesis here to note that this exhibiting function of giving behavior is not the only reading we could make of it. I was able to observe one scene in the course of which the gift signal—which is easily recognizable—was given by a helper at the nest without food being given to the chicks. The helper was not visible to other adults—the nest was camouflaged by a dense growth—but was perfectly audible. In contrast to the Zahavian hypothesis of exhibition that I proposed to him, Jon, the Oxford zoologist[12] whom I accompanied that day, posited an alternative: the bird did not emit a signal aimed at the group, but rather assessed the real need of the brood in measuring the intensity of the response to its proposition. We will return to the particularity of this hypothesis. For those interested in how the frame of reading can affect the gathering of data and their interpretation, it is necessary to remark that this explanation is in line with the sociobiological theories of Robert Trivers. They describe the cooperative relation between parents and children as one potentially presenting a conflict of interest: it concerns a veritable war between the generations in the course of which, according to Trivers, the children will miss no opportunity to cheat. They will pretend, for instance, that they are hungrier than they actually are. It is therefore in the interest of the parents to measure, to the extent possible, the real need of the children. Nor will we be astonished to observe that it is the general model of the "arms race" (Oxfordian, according to Zahavi) that gives its meaning to this hypothesis for reading behavior: we find ourselves in effect confronted by the behavior of the "liar," that is a parasitic behavior, which is met by a "mind reader" behavior that plays, in some sense, the host and tries to counter the "manipulations" directed against it.

We close this parenthesis by returning to the Zahavian reading of this series of anecdotes. If gifts or relief of the sentinel role are occasions for conflict when it is the subordinates who offer them, and if the behaviors called altruistic are carried out in a rather univocal manner from the dominants to the subordinates, it is because they have a function in relation to the hierarchy that they seem to indicate. Cooperation must be a privilege, the sign of a status, and this is rightly so, since cooperation is the sign of access to resources, of availability in terms of time, force, and energy. If giving is a privilege, and if to give it is necessary to have, Zahavi concludes from this that the gift must have an exhibition function. It can thus constitute a signal.

In his first articles, Zahavi attributed the finality of this type of signal to the establishment and maintenance of the hierarchy. The later articles will establish a distinction between hierarchical status (dependent on age) and social status (dependent on individual performances).[13] A few days after hatching, the chicks of the same brood will determine rank in the hierarchy through conflicts. These conflicts will continue to diminish until becoming extremely rare as adults. In simple social structures, an elevated hierarchical rank will give an individual the power to get just about anything it wants from subordinates. But, in more complex groups like the cooperative groups of babblers, the members must establish compromises, and even resolve conflicts—for example, around the fact of determining who has the right to mate. Status can thus nuance the hierarchical relation and favor individuals of an inferior rank in the management of compromises around this right. The possession of an elevated status, faced with a dominant individual in the hierarchy, allows the hierarchically dominated individual to obtain concessions from the dominant one. In the same way, when two babblers are similar in status—for instance, because they are part of the same brood—status can determine the degree of dominance, that is to say the power that one will have over the other (power being defined by Zahavi as constituting the access to resources).

Altruism, as a performative exhibition of status, will be, from this perspective, the mode of conciliation between two contradictory pressures that we identified as constituting the problem to grasp: it is the perfect compromise between the need to cooperate and the need to affirm one's superiority in the competition around the definition of this status.

The stability of such a system cannot however be assured unless potential cheaters do not have the possibility of using "false signals": the *handicap principle* will be the necessary corollary to a system of this kind. The handicap is defined as being the cost of each of these signals that expresses the value of an individual. It is the quantity of energy, the risk run, the quantity of resources utilized, the burden that each of these "publicity" behaviors represents. In this, it is the sign that the individual has the means to take on such an expenditure in terms of energy, risks, and resources.

The cost, on this account, does not come down to being a simple consequence of the signal, although it is certainly an element of it, an essentially reliable element of the system of communication that indicates the value of the individual. In this way, if the altruistic behavior is the optimal compromise between the necessity of preserving the availability and physical integrity of the partners and the necessity of facilitating one's own access to resources, it at the same time is a reliable indicator of the qualities of the male in the eyes of females.

This reading of behaviors in terms of what I call "performative exhibition"[14] of the social status had already been suggested in the literature treating social behaviors among humans: certain theories of potlatch could be connected to this. The interpretation of the gift among infants proposed by Susan Isaacs refers to the feeling of power experienced by the infant from the very fact of being strong enough as not to be selfish. Walter Goldschmidt thinks that the resolution of conflicts between the interest of the group and the interest of the individual in primitive societies only takes place thanks to the particular involvement of the participants under the form of an elementary need for recognition, approval, esteem, dominance, and so on.

Some of Lorenz's writings on the jackdaw could be reread in light of this type of explanation.

The animal photographers Jim and Kathy Bricker were surprised by the bravery shown by certain ravens they photographed. This bravery was all the more shocking since the ravens stole food from wolves gathered around a carcass, and they lost interest in the food once the wolves were gone. If the pilfering was not correlated to hunger, how can we understand such audacity? Bernd Heinrich put forward the hypothesis that this type of action would be destined to improve the social status of those who produced it.[15] Along the same line of thinking, Trygve Slagsvold showed that placing a decoy of a predator on the ground in front of the ravens creates a competition between males for the privilege of attacking the decoy, and they generally shoo away the possible co-attackers.[16] Some of the attacks toward the other "mobbers" can even lead to the death of congeners and a total disinterest for the decoy. I would however raise some reservations regarding the value of this last datum: if the decoy were not one, if it were instead a real predator, we could think that the disinterest toward it would have consequences such that it could not be maintained very long.[17]

These theories received a somewhat limited reception to the extent that they went against general theories explaining gathering or "mobbing."[18]

THE HANDICAP PRINCIPLE AND SEXUAL SELECTION

We have indicated above that altruistic behaviors constitute not only an optimal compromise between contradictory pressures (1), a performative exhibition of status (2), but also a means of drawing the attention of females (3). In this last sense altruistic behaviors can be considered as sexually attractive characteristics. This investment by males with the goal of drawing the attention of females is comparable to the handicap taken on by competitors in order to bolster the credibility of their claims to superiority. Extravagance becomes, from this perspective, the constitutive element of the signal itself.

Classical theories that would attribute to extravagance a role of preferential stimulus, even supernormal stimulus in the choice of females, cannot, as we recall, explain how it simultaneously plays a role in the intimidation of rivals. This difficulty led the intellectual heirs of Darwin to consider sexual selection as a process that is separate from natural selection and sometimes even in contradiction with it. In effect, the signals used in sexual selection do not seem to obey the same rules of efficiency and economy: on the contrary, and we have already made allusion to this, based on their extravagance one could consider these "productions of bizarre forms" to be "wastage," or "non-sense" that is useless because it does not benefit the species.[19]

WASTAGE AND RELIABILITY

With the handicap principle, the wastage itself defines the signal. Remember that we are here within an optic according to which every cooperative relation potentially implies a conflict of interest: in this way, when a female must choose a male, her interests are not necessarily those of the male. The male, for his part, will act with the goal of being chosen, while she must choose the male who will assure her the best possibility of reproductive success. Under these conditions, the female must not respond to non-reliable signals.[20] If one considers the variety of characteristics imposed by sexual selection, one notices that almost all of them represent a handicap for survival. Now, the animal who is the bearer of them has survived. The handicap will be like a test imposed on the individual: a female confronted with two males, one of whom carries a heavy handicap, and the other a less encumbering or less visible burden, will choose for her best interests—in terms of reproductive success—the more handicapped of the two, because his handicap is the reliable signal that he has passed the test. The handicap is thus a reliable signal borne by the male. If there were no link between the secondary sexual characteristics and the real quality of the one who bears them, "the door would be open to bluffing."[21] The one who receives the signal will act in her best interests if she can verify the reliability of the information encoded in the signals she receives, and will not respond to signals that do not seem to her to be sufficiently reliable.

One can understand, from this, that the handicap considered in itself will not only be the reliable signal of the quality of the male to the female, but equally toward rivals as well. And in this case, too, the signal is reliable, since the cost of its utilization is such that it surpasses the potential gain for an animal who would cheat. Going out into the full light of day displaying bright colors or noisy vocal displays is risky, because the colors and the noise cannot fail to attract predators or potential rivals. The individual who has survived this type of strategy indicates, by the strategy itself, that he has the means to take it on. This constitutes the cost of the signal, its reliable element. The song of a bird is a reliable indicator, by its duration and intensity, of its state of health and the availability of resources at its disposition.[22] A heavy antler or an encumbering tail constitutes a reliable sign of the physical vigor of the one who carries such a burden.

In this respect the handicap principle establishes a logical connection between particular models of the signal and the message encoded in it.[23] This means that for each message there is an optimal signal that best amplifies the asymmetry between the "honest" signaler and the "cheater."[24]

THE EVOLUTION OF SIGNALS

The importance of cost in the evolution of signals is such that they can no longer work if their cost is eliminated or significantly reduced. In a territory infested with predators, vivid colors can play the role of a signal for the ability to escape predation. The augmentation of risk is compensated by the benefits (in terms of reproductive success) drawn from the fact that the animal exhibits its capacity to escape these risks.[25]

If the species migrates toward a space devoid of predators, contrary to the predictions of Lorenz about intra-specific selection acting alone, the colors will lose their value of advertising this capacity and thus soften. This seems to be the case in the example given by David Lack: ducks living on islands seem to be less vividly colored than the related species on the mainland. Ernst Mayr suggested that this loss of colors was the consequence of the decrease of the risk of hybridization, since insulated populations of ducks are no longer in contact with closely related species. The theory of signals proposes an alternative hypothesis: on the mainland, risks of predation played the role of selective pressures in favor of colorful plumage, because the vibrant colors are a signal of the quality of the individuals who carry them. The absence of predators in insular conditions led to the disappearance of this selective pressure and the characteristics that it favored.

Colors, as specific ensembles of decoration, were generally interpreted as conventional signals allowing the identification of the species, the age, or the gender of the individual. According to Zahavi, this role would be a secondary one, conventional signals representing instead the model or standard allowing each individual to infer, based on the variations that these standards show, the qualities of others. The decorations are thus messages that carry reliable information regarding certain characteristics of the individual: a point in the middle of a circle shows its circularity, a median line allows the detection of asymmetries.[26] The decorations play a role in the amplification of differences and constitute a source of information for the potential partner in a relation. In the same way that rituals constitute a common model permitting expression—and being informed about—individual variations, conventional decorations are the indicator of differential quality, and will be selected as reliable signals.

Altruistic acts would thus be considered from this point of view as behaviors functioning according to the handicap principle. They are the sign that the individual "has the means": giving food is a reliable signal of access to resources, informing someone of a danger is a way of displaying the fact of taking on certain risks. Altruistic behaviors thus constitute handicaps by which an individual claims its superiority over others, and the cooperative system will above all be defined as a highly competitive system.

Somewhat paradoxically, the theory of signals will insist, among other things, on the common interest that can exist between a prey and a predator. This paradox disappears, however, if we consider the theory of signal selection within the framework of an interaction: the signal evolves as the result of the common interest of the parties, even if they are in conflict, and will not be selected unless they favor the transmitter and the receiver. The receiver, in effect, avoids reacting to false signals, which generally insures the evolution of reliable signals.

A prey population is composed of different phenotypes, some of which are more vulnerable than others. If a predator tries to catch a less vulnerable prey, it loses time and energy. It also reveals a presence to other potential prey. In sum, it risks coming home empty-handed. Prey who uselessly flee also lose time and energy. It is therefore "in the interest of both, the predator and the non-vulnerable prey, that the latter identify itself clearly to the predator."[27] This identification must be reliable to be selected—the signal must be perceived in such a way that it allows the predator to classify its prey as a function of its vulnerability. Likewise the bounds (stotting) of the gazelle, elsewhere interpreted as an act of altruism for the attention of congeners, will be considered, in this theory, as a signal that is in fact addressed to the lion: the ostentatious leap seems, according to Richard Dawkins, in Zahavi's commentary to constitute a real provocation with regard to the lion. In fact, it testifies to the vigor and health of the gazelle and, for the lion, constitutes the reliable indicator of its prey's invulnerability, as if it told him (as supporting evidence), in the Dawkinsian style, "Look how high I can jump, I am obviously such a fit and healthy gazelle, you can't catch me, you would be much wiser to try and catch my neighbor who is not jumping so high!"[28]

These exchanges of information, which are signals, thus seem to largely surpass the strict frame of reproduction so as to extend to all the domains of relational life for living organisms.[29]

Revisiting the distinction from Darwin, Zahavi posits that there are indeed two distinct selective processes. But to take account of the extravagance of the transmission of certain information—both within and outside the frame of reproduction—it is rather necessary to invoke a vaster mechanism than that of sexual selection, and of which it would be only a subset: the mechanism of signal selection.

4. Models and Methods
Outline of a Field Study

A difficult question emerged from reading one of Zahavi's articles devoted to babbler behavior: were these birds extraordinary due to their various dances and games, as well as their disputes and competitions in order to provide the privilege of being altruistic, or was it rather Zahavi's singular gaze that ultimately conferred originality on these birds?

After reading this article, my questions had to change.[1] Indeed, the literature I surveyed at the time was generally composed of articles describing, in a relatively sober manner, automaton species without creativity, presenting rather conventional behaviors in this context. To answer these new questions I had to visit the site myself, pose questions, read, and observe.

I will not leave you in suspense, and everything that precedes this allows you to predict my response. Deceptive, no doubt, but in any case clear to those who read me, the solution appeared to me as wonderful in its complexity, because it preserved the two poles of my question it did not reduce one to the other, nor did it exhaust any of the meanings. Babblers are extraordinary birds, and Zahavi's theories emerge in rather unusual contexts of justification.

EQUILIBRIUM AND THE GREAT DIVIDE

An ironist would tell me that the birds appeared to be extraordinary because I saw them through Zahavi's eyes. I was myself this ironist when I arrived at the Center. I was correct—the ironist was right—but it is necessary to go beyond this con-structivism that is unfair to the "nature" pole: not all birds share this same tendency for dancing, nor do they maintain such a complex social structure as the babblers. One obvious reason for this is that not all birds live for fifteen years.

It is also rare for birds to let other creatures get in such close proximity to them. Let us stop here, just for a moment, and reflect on what has just happened: without noticing it, we have just slid from the bird to the observer, and have as a result placed ourselves within the space that unites them. It is here, in this space that is both geographic and relational, that the double question can emerge. The essential and pressing duplicity of this question is what transforms all research into ethical, aesthetic, and ethological research: "Who am I, how is my gaze structured in order for you to appear to me such as you are?" together with "Who are you such that I see you this way?" Asking the first question without the second leads us to a sterile constructivism, whereas asking the second ques-tion without the first leads to a dogmatic realism. Together, they form an instant of equilibrium, the best moment to think things in terms of relational and complex dynamics. These moments of equilibrium in space that bring together natural objects and the questions that investigate them are the mo-ments, no doubt, that contribute to effacing the great modern division between nature and culture. And with this efface-ment, the great divide between the human and the animal finds itself seriously called into question: Isabelle Stengers[2] sees the signs of this questioning in the heresies of the new primatologists (such as those of Shirley Strum). In leaving the baboons to respond to questions other than the ones tradition-ally posed to nonhumans, in according to them, for example, social abilities that would not be the simple product of obedi-ence to specific rules, but the outcome of their creativity in the construction of social ties, these primatologists seem to me to outline the structures of a space of equilibrium. What has happened there such that the old questions are abandoned in order to formulate new ones? The answer that Stengers suggests leads us to another point of integration at the core of our problem, to what we have said about the ritual and dance: the primatologists knew to abandon the research on invariants that individuals obey, in the same way that ritual and dance led Zahavi to question, beyond specific similarities, individual singularity. If we linger for some moments over Shirley Strum's research—over what Isabelle Stengers calls her *quest for pertinence,* such as it is related both in her book *Almost Human* and in the article written in collaboration with Bruno Latour[3]—we perceive a parallelism of questions with those that are posed here for the babblers: when she is confronted with surprising divergences between what she sees among the baboons and what the data of her colleagues and the literature in general had prepared her to see, Shirley Strum concludes that each troop of baboons deviates from the norm. But this conclusion is not, one might suspect, without its own problems. In the attempt to find "a way out of this dilemma of intra-species variability," Shirley Strum considers two possibilities: she could, in the first one, "reject data and the views of the observers. A common position was this: other baboons did not behave differently, they were just inaccurately studied."[4] Another way out would be—and it is the position that she and Bruno Latour propose to adopt—to call into question both the epistemology and the ontology underlying the research, to place herself in a position that takes account, simultaneously, of the observing subject, of her modes of knowledge and definition of the object (the paradigm), and of the object itself, become subject in its turn: "The traditional, ostensive definition of baboon society has been unable to accommodate the variety of data on baboon social life. As a result, some information has been treated as 'data' and other information as discrepancies to be ignored."[5] Considering, with the performative paradigm, that "baboons 'perform' their society might also allow a more consistent interpretation of the cross-populational data and data from other species of monkeys and apes."[6] As is the case in numerous controversies in ethology, it is not simply a question of admitting that the capacities of the observer can, by themselves, explain the deviations from the norm, but a question of enlarging the paradigms, with the goal of making them more supple, more open in the face of difference, to the variety of contradictions seen. Substituting the performative paradigm for the paradigm of exteriority at the epistemological level leads to two interesting shifts at the level of the ontology coming out of the research: other "beings" acquire the right to existence, the objects become subjects. This is also where we can see another space open up among primatologists, what we have called a space of equilibrium between the pole of subjectivity and that of objectivity, of nature and of culture, of the gaze and the object observed.

In the same way that the baboon was able to necessitate that the primatologists change their questions, the babbler offered Zahavi the opportunity to pose questions differently. Here we situate our response, in a slightly artificial manner, on the side of the babbler, an active subject of the relationship that brings together the researcher and his object of study. The space of equilibrium is, rather, situated in the relationship itself that unifies an active subject and an observer who is also active, in the game of questions and responses that characterize the bond created between them. This space of equilibrium can, I believe, partially be described in the methodologies: those that, indeed, give a frame and tools to questions as well as to answers.

A PRIORI AND A POSTERIORI METHODOLOGIES

When one analyzes the relation between the concepts that underlie the search for specific similarities and the methodologies utilized to observe them, a common term appears for the object, its categorizations, and the way of going into the field: a priori. The invariants of behavior are, in some way, the a priori of the program and the innate mechanisms of release: does

Lorenz not declare himself the heir of Kant when he maintains that the organization of behavior is a priori?[7]

To these a priori correspond some methodologies that can be called *a priorist* even if they are neither the consequence nor the condition of it. The general characteristics of the method can nevertheless facilitate the perception of invariants. The researcher equipped with an a priorist methodology goes into the field with a hypothesis to which he intends to submit the facts (here the term "submission" deserves to retain its full ambiguity). This methodology takes its authority from prior research, and organizes the work according to hypotheses elaborated in those theories. For example, Jon, the Oxford researcher, has great difficulty seeing how one could possibly go into the field without wanting to test a hypothesis.

The a priorist will thus research the responses to "how" questions. To be precise, his trump card will be the regularity of the real: invariants can establish this sought-after regularity. So, in a certain way, the approach and the pole of interest will find one another in a system of resonances, a mirror system.

The a priorist will more willingly use the experimental approach: the manipulation of variables—to understand the real in resisting it (according to the words of Gaston Bachelard)—will be for him the right method to measure and understand invariants. In this way the method of the decoy wonderfully illustrates what we have just been saying. For a bird's egg, one can substitute a larger egg, a smaller one, a redder one, or whatever else, that allows the variations of the invariants, that is to say the instinctive or preprogrammed response of the bird in the face of the stimulus signal that some characteristics of the egg represent. The a priorist, in the face of variety, will try to create variation in shifting away from ordinary conditions of observation. The a priorist, in a certain fashion, imposes his question and the limits of the response on the real, as the program imposes its questions (the stimuli signals acting like a key in a lock) and the behaviors that respond to them (the innate responses). In the same way, one can also note that the methodologies and the objects enter into resonance and create an isomorphism, forming a mirror effect.

If we envisage the approaches on a continuum, we find on the other end from the a priorist approach that of the *a posteriorist*.[8] Zahavi's approach illustrates it well. The term's reference to "what is known after the experience" relates here not to the experience of the experimenter, but to the most common experience.[9] Before going into the field, no hypothesis is formulated in an explicit manner. It is rather a matter of going there with the sole intention of seeing what will happen first, and then of establishing hypotheses and interpretations regarding what one has seen a posteriori, as a second step. Certainly, hypotheses are not totally absent, but they are generally implicit, and go beyond the frame of the hypothesis stipulating that there is something to see. With Zahavi, one could, for example, think that the influence of the theory of individual selection is accompanied by some statements of uncertainty.

This approach resembles, from many methodological and theoretical points of view, the anthropological method. The a posteriorist collects anecdotal events and tries to give them a meaning in creating links between these happenings. Rather than variation, it is variety that she engages, and on it that she bases her experience as a type of common experience. Zahavi critiques the game theory of evolution, for example, because it does not take account of the fact that animals in the field "react in a highly variable way . . . Reaction seems to be determined by information gathered rather than by a pre-set program activated by simple arbitrary signals."[10]

Of course, this polarization is an artifice, and we practically never encounter a totally pure approach. The clearest illustration of these two approaches is found in Lorenz, who can pass, in the course of the same research, from one extreme to the other of the continuum. Even though he is animated by an a priori theory—the behavioral invariants of the innate program—Lorenz can adopt, in the field, the double approach: the anthropological approach as well as the a priorist, experimental approach of the research into invariants. His research on imprinting during the following reaction among geese constitutes a striking example of this hybridization of approaches and objects. To recap, the theory relates the fact that a young bird, some hours after its birth, experiences a critical period in the course of which it will follow any nearby moving object.

The theory itself bears the mark of the hybridization of methods, since it interprets the imprint as resulting from the conjugation of an innate program that is invariant in its form, and mechanisms of learning open as to their object. The hybridization of the theory thus faithfully reflects that of the objects: the experiment of manipulation of variables at play is carried out in two frames. As an experimental approach, it verifies its hypothesis and demonstrates the innate response, a priori, of the bird; as an anthropological approach, it shows what is acquired by the bird "by experience," what goes beyond the frame of invariants, what we could call then the a posteriori of behavior. This approach is all the more anthropological in that Lorenz himself plays the role of lure, subsequently using, in a quasi-anthropological manner, the relation established with the subject of his experiment: the young goose who takes him for its mother.

The two conjoint poles of objects and methodologies are reflected this way in this particular operation: the experimental approach shows the a prioris of the program (the innate following response); the a posteriori or anthropological approach, for its part, shows the variety that the program did not predict. It allows a space for surprise and creativity to emerge.

THE INVESTIGATION AND THE TRIAL

Other differences can appear here or there in our continuum and give each approach its own characteristics. These appear in what one could call the procedures for research and the construction of theories.

The a priorist, of whom we said that he fixes the real in the strict frameworks of his answer, does not believe in what the real is telling him: the duplicity of causes, the obviousness of bias, provide so many occasions to think badly. He will force the real to restart the story again and again. He expects of it that it does not contradict itself, that its versions be correct, that the regularity of the shifts in relation to the original response (that obtained without the lure) confirms his hypothesis. It is not only the real that is put on trial, but he himself in his relation with the objects (witnesses) and the hypotheses describing a system of causes and effects. Testing the real revisits the etymology that designates the procedure: it is a matter of putting one's witness to the test, as in a judicial process.

Altruism, studied by the a posteriorist, must, as far as she is concerned, make up the object of a meticulous study: it is necessary to collect clues that contradict the innocence of the behaviors, to search out the meaning beyond the appearances, to compare apparently unrelated events. To the *trial* of the experimenter corresponds, then, the *investigation* of the a posteriorist.

This investigation that engages variety rather than variation is, however, not without its own difficulties. Indeed, how to make any explanatory interpretation based on it? If the process of variation is reduced to the confrontation of fictions to describe it, the investigation into variety will reveal neither meanings nor criteria that allow choosing between these fictions. The hypothesis results from juxtaposing heterogenous observations that only make sense once they are collected together, but it cannot unite opinion except on the principle of adhesion to a fiction.

I called this manner of proceeding "investigation," not only because it is situated at the moment before and is situated differently from the trial, but also because it bears the traits and characteristics of the police inquiry as told by the masters of suspense: the researcher not only places heterogenous events side by side, but most importantly brings out the facts "that do not stick" with certain of the witnesses, or "that do not hold up" between them. Recall the manner in which the chronology of the theory unfolds and how I presented it: to the question of knowing how such an altruistic bird did not convince Zahavi of the pertinence of the theory of group selection, the answer was in some sense that Zahavi does not believe in the "innocence" of the babblers. And for good reason: there are too many troubling elements, too many occurrences that "do not hold together" and that, once placed side by side, start to make sense and to describe what is really going on (according to the hypothesis where Zahavi is right, of course): the help at the nest is more harmful than beneficial; to a soft hold of the hand corresponds an aggressive behavior and vice versa; the reprimand from the dominants to a congener who interferes takes the strange forms of gift-giving and preening; not everyone can offer gifts in the group; some gifts, but not all, lead to conflicts; dances always take place "when one should not, where one should not." This meticulous collection of small pieces of evidence that shouldn't be found alongside each other, or that seem to be expressing something other than what they indicate on the surface—is this not exactly like a police investigation?

So, this investigation, which is itself a fiction—a fictional and hypothetical construction—has no other power to convince than by way of what it suggests. And this fictional investigation becomes the very response to the entire process of experimentation as a confrontation with all the myriad fictions. For a long time, this was one of the reasons for the indifference or hostility toward Zahavi's theory in scientific quarters.

The trial over the mode of variation offers, for its part, the dispositive of confrontation for these fictions, which is to say the experimental dispositive that, as Stengers[11] defines it, tests them and allows deciding between them. A fiction that can oust the others can take with it the adherence of other researchers. We will perhaps be more concrete in visiting the sites for the testing of ethological fictions during the trial. This description will allow us to scrutinize all the differences between the two approaches.

EXPERIMENTATION AS PLACE OF TRIAL

The manipulation of the real is a way of verifying the "how" of what one has observed: "what is at work here, what is the cause of what I have observed?" The manipulatory procedure will consist of a modification of the real in order to accelerate it, vary it, and insert it into a series of constraints with the goal that it will meet laboratory conditions while also respecting the conditions of nature.[12] But in the procedure itself, nothing indicates to us what will play the role of guarantee that we are speaking of the correct "how," of the guarantee that what takes place is indeed the effect of the given cause, that the fiction is the true fiction.

The way the proof is organized will play the role of guarantee, because this organization will, in a gradual manner, put each of the fictions to the test of the other fictions, as in a trial. Observe how an experiment is organized: Jon participated in the research of a group of Oxford zoologists led by John Krebs that, in collaboration with a group in Toronto led by David Sherry and Sara Shettleworth, tried to elucidate the manner in which some birds (the American black-capped chickadee and the European swamp chickadee) memorize the hundreds of hiding places for food that they use.[13] The question here is clear and is expressed simply in terms of a "how": what do they do to relocate the caches? The initial hypothesis—we are following an a priorist approach—is that the birds use their memory, and not olfactory indices or other processes of association with non-mnemonic indices.

Starting from this hypothesis the approach will unfold in a discursive manner. The progressive elimination of alternative fictions will itself take on the narrative allure of a fiction or an imaginary trial. In the field, the researcher is alone—or with his assistants, close collaborators, or students—but he will assemble around him the imaginary personages of a trial. Each of the imaginary personages seems to interpellate the researcher by posing an objection to him in the form of another fiction. The imaginary trial will progress this way until each of the fictions proposed by these personages is eliminated. Let us follow the steps: in an artificial forest, one of the researchers dug a hole in each of the trees, then covered it over and sealed it with Velcro. We know, the author tells us, that the bird can open the Velcro covering since it resembles certain natural conditions of the closure of orifices. You will see that even before the experiment has begun, imaginary objections already start to intervene. The Velcro is for its part used to ward off another objection: the bird would not have to prove its memory if the food were visually accessible.

In the first stage of the trial, the birds hide the grain, and are then taken from the forest for twenty-four hours after which they will be rereleased into the forest. They are able to rediscover the little caches and the grains they contain, and thus allow the researcher to attribute the success of their performance to memory.

Another objection seems to have been created since our researcher starts the experiment again, but with the addition of a variation to address the possibility that the birds could find the grain thanks to its smell. The birds could also trust in the fact that the Velcro does not cover the caches that had been visited in exactly the same way. The researcher will take away the grain after the first visit in the absence of the birds, refasten the Velcro, note where the grain was hidden, then, after twenty-four hours, observe the return of the birds. They go right for the caches used previously.

Once again another imaginary objection comes up: "The birds do not memorize the hiding spots, but use certain indicators or criteria, always the same, to hide the grain. Rediscovering the caches is therefore not an exercise of memory, but an association of habits similar to those that we generally use to park and find our car." The researcher thus hides the grain herself and shows the caches to the birds. This had no consequence on their performance.

After having thus eliminated all the imaginary objections, in sum the rival fictions, the researcher finally has the chance to demonstrate that not only is her hypothesis correct, but that it is perhaps even better than what the fictions have shown. An experiment in two stages puts the chickadees back into the presence of the grain to let them eat half of it. In the second stage, twenty-four hours after the first phase of consumption, forty-eight hours after she had hidden the grain, they were once again released into the artificial forest. The chickadees will show that they have not only held the caches in their memory, but that they have, beyond that, memorized the caches already visited in the course of the previous passage. They consequently remember not only the location of the full caches, but also the empty ones.

This testing in an imaginary tribunal seems to constitute a guarantee against subjectivity, against the opinion of one individual. The experiment becomes the site of an imaginary public space, a fictional way of opening the doors to the natural laboratory. This testing in some ways plays the role of the trial in the first instance, the publication permitting, for its part, the procedure of appeal for unjustly discarded fictions or for sanction of fictions with exorbitant pretensions. During this stage the imaginary jury ceases to be so in order to fulfill the role of a real jury, that will not miss any of the overlooked objections or any of the defects of the procedure having taken place in its absence.

THE BABBLERS ON TRIAL

If we now turn toward the trial in the course of which Jon will organize the babblers, we no longer find all the characteristics of the procedure. Indeed, this trial is still in its first phase, during which the possible alternatives are considered. The trial itself will not start until the following spring. Nevertheless, we can already notice some of the hallmarks of the confrontation of fictions in the tribunal of experiment. Here the alternative fictions concern the explanatory hypotheses of helping at the nest: initially, the first hypothesis—coming out of the theory of kin selection—the birds help their relatives; the second fiction sees helping at the nest as a system of exchanges founded on reciprocity. Jon's last fiction is the Zahavian hypothesis: the birds are "helpers at the nest" because this represents for them a good means of raising and demonstrating their status. How to decide between these fictions? It is necessary to find an element of variation for each of them in order to make the situation amenable to experimentation. The form the help takes seems to meet the required conditions, and one can make it vary by introducing decoys. The way of feeding the young seems to reflect the conflict of interests in the group: for example, the kin theory predicts that the help will be positively correlated with the degree of relationship but negatively correlated with the size of the group of helpers. This negative correlation is rather logical. The more helpers there are, the less the necessity to work for the brood. On the contrary, within the frame of the Zahavian theory, the size of the group does not have to affect the help, since it is useful not to the recipients, but to the giver. The fiction of reciprocity, for its part, must be evaluated on the basis of observation: do current helping relations reflect previous relations and do they permit the prediction of future relations? It would seem that the observations carried out up until now do not offer evidence in favor of this fiction.

A first stage of the work consists in observing, during a certain period, the way in which the natural variations affect the relations of helping at the nest. During a second stage, Jon will try to produce variations that would have to allow choosing between the fictions: if the effort of a member of the group increases considerably, one can predict, as a function of each of these hypotheses, different consequences for each variation produced. In this way, if it is the fiction of kin selection that is the "true" fiction, then in response to the increase of the efforts of one individual, the other helpers will diminish their efforts. It is not necessary to force-feed the chicks; if they are fed, all of the helpers can occupy themselves with other things. On the contrary, if Zahavi's fiction is true, and if the act of feeding the chicks constitutes a demonstration, then even if an individual augments its services, the other helpers will maintain their level of effort, or perhaps better still will enter into competition and augment it. For this first stage, Jon will proceed to a simple manipulation: he will place artificial food at the disposition of some of the helpers that will allow them to easily augment their apparent investment in alloparental help.

During another stage of the process, Jon also plans to intensify the pressure on some helpers. Three solutions were considered: attaching a weight to the tail of some of the babblers; reducing the mobility of the beak of some of them (by a system of small attachments); or else producing a decoy: the cries of the chicks will be recorded and played near the nest, during the passage of certain selected individuals, with the goal of inducing, among them, the belief in a very intense demand on the part of the brood. If they augment their efforts, we can measure the diminution, the stabilization, or the augmentation of the efforts of other members of the group. This possibility must be, in principle, the only one used—for clear ethical reasons.

If the imaginary tribunal of the experiment guarantees the a priorist approach, the question must now be turned toward the a posteriori to ask what guarantees this method.

APPROACH AND POSTERIORITY

Contrary to the experimental procedure creating an identity between what founds the hypothesis and what assures its validity by means of decisive experiment (showing "how it works," and at the same time giving the clues that "it is like this that this works"), Zahavi's approach separates the two moments of the procedure. Zahavi himself separates them so well that he himself never attends to the testing and leaves the care of this step to others.[14]

Indeed, the tests that the theory underwent were carried out elsewhere, and for the most part contradict the theory: Robert Slotow, Joe Alcock, and Steven Rothstein, for example, change the colors of the feathers of subordinate sparrows to disguise them as dominants.[15] According to these researchers, if the theory of social control of status is correct and "cheating" is impossible, then the subordinate usurpers would have to undergo a change in the intensity of aggression—that is, they'd have to become aggressive more often. This was not the case. Since the usurpers were not unmasked, the authors were able to conclude that the principle of reliability, which is at the base of the handicap principle, was a false fiction. Fictional laboratories also put Zahavi's theory to the test. However, after having refuted it, the procedures will show that the fiction was realistic. We will come back to this later.

The approach of collecting singular and anecdotal indices that make up the preparatory work for the investigation tries to find, beyond and despite the invariants and the a priori, the particularity of the response for each of the individuals. We have made reference to the heresies of the primatologists, and we find them again here in the daily confrontation between the researcher and his babblers. The heretical approaches of the former (the primatologists) as of the latter (Zahavi) are, in fact, approaches used by anthropologists. And it is here too, in my opinion, that the great divide—so pointedly scrutinized by the new questions posed by the primatologists—experiences the signs of its erasure.

This erasure can be explained as much by the methodology and approaches used as by the particular personality of the bird. Zahavi is not only accepted in proximity to the group, but integrated into it, and always moves in the milieu of the birds (not as a stranger observer would do, from the outside, in front or behind). Most of the birds observed are tame—the others are called barbarians. Each bird receives a name so as to help in later identifying it: the birds, like all the living beings that draw away from us, appear too similar to each other for us to be able to definitively recognize them; each bird will also be identified and named thanks to four colors of band placed around the leg of each bird: the name of each one is thus formed from four initials of the colors in Hebrew—in this way, when a researcher says they have seen MMCT or AMMT, everyone knows exactly who they mean, and to which group they belong.

NEW TIMES, NEW ETHOLOGICAL SPACE

Zahavi, like an anthropologist, learned to recognize each of the individuals and to know, over the years, the evolution of its alliances, its status. Thus one can find out, in consulting the figures and the archives on the big board permanently displayed in the laboratory (and some articles that adopt this narrative form), that AMMT was born in 1975, that he lived with his father and one of his brothers from the same clutch. Another female replaced his mother, in 1977, and he became a "breeding helper" (reproductive helper at the nest). His dominant brother was chased from the territory and AMMT could thus, with the death of his father, become the dominant male. In 1979, he sired five chicks, and six between 1980 and 1981. In 1982, there was once again a change of females and he could not breed with her. The two sons that he had with the departed female became breeding helpers. He chased one away and remained with the other up until his death, in 1987. AMMT thus remained in the same territory all his life, waited to become dominant, lived twelve years, and had twenty-nine children in the end.

In a similar manner, we can also explore the more tumultuous story of SMTA, who had to conquer a new territory and lose his collaborators, or again that of MTMC, who was chased from his territory by his brother. He lived as a refugee, before being rejoined by two of his sons. After some years of occupying marginal territories, he was able to replace a male who had just died, and form a new group. The time of a bird is in this way structured like that of a human, and the stories become the stories of a life. An anthropological time frame is thus substituted for an ethological time frame.

Space, too, is structured in a particular manner, since Zahavi observes from amidst the birds. The ritual of approach and meeting prepares—and no doubt defines—the relation: Zahavi whistles in imitating the call of the babblers and he throws them some bread. The babblers respond by drawing near and jumping, eating the bread, and then resuming their activities without seeming to be too preoccupied by his presence. The particular structuring of space is possible thanks to the familiarity of the birds. However, while this structuring of space is possible, it is not necessary, and each of the researchers integrates the relation in a markedly different space: Jon, the zoologist from Oxford, experimenter in constant trial with the fictions, establishes the relation of a distant spectator with the birds. He is either before them or behind, never among them, never too close. Osztreiher, the doctoral student assistant, for his part, is always in front of them, but in a way that is so close that the other researchers reproach him for perturbing the birds (one day, in order to observe the birds more closely, we saw him accidentally block with his head the entrance to a bush where a nest was located).

The spatial structuring of the meeting between Zahavi and the birds is drawn as a space without limits between the observer and the observed. In an analogical manner, a common mental space can be seen as corresponding to this limitless space, that is a common mental space of anthropomorphic identification: "When we don't understand why an animal does something one way or another, we try to understand what purpose such a behavior could serve for humans, then we try to see if the hypothesis can be confirmed by these observations."[16]

Zahavi calls this procedure the anthropomorphic model.

This model must not be understood here as a model in the technical sense of the term, but rather as a psychological model, a referent of identification. While being a contested heuristic strategy, it was nonetheless the tried and tested method of many hunters: the bushmen, for example, identify so much with the animal they are in the act of hunting that they can respond to questions like "What would I do now if I were the animal?" with a prodigious exactitude.[17] Zahavi's approach turns on an essential point: interpretation is substituted for prediction. The question is not so much "What will it do?"—a question that could confirm or disprove the fiction—but, rather, "Why does it do it?" a question for which there is no answer that would have the power of striking down any alternative fictions. Interpretations always suffer from the impossibility of consensus on their subject. In other words, as I explained regarding the investigation, the sharing of fictions can only be founded on shared opinion: discord expresses the incapacity of the interpretive statement to be anything other than a simple fiction dependent on the intentions and convictions of its author.[18]

This incapacity is all the more marked since the anthropomorphism is derived from an identification of causes of behavior with the intentions of the animal who exhibits it. In sum, we have clearly understood from the very beginning that the "how" is much more of a "why." The fictions encounter and confront one another without allowing for a choice to be made between them. This decision work will be done elsewhere, and we will speak of it later.

Anthropomorphism and the anthropological method accompany a fictionalization process of a very particular type: not only does every action the animal takes receive a meaning, but each of its behavioral sequences must be understood, according to Zahavi, as the best possible compromise by which the animal does what it can do.

This process of saturation of meanings and the elaboration of positive and valorizing fictions is not merely tied to the attributive or projective mechanisms of anthropomorphism nor to the anthropological approaches characterized by spatial proximity and personal relationships, even if it is largely dependent on them. It is founded on an epistemology of evolution that doubles as an ontology of nature: "The people from Oxford argue for the stupidity of the host of the cuckoo. For me, if the bird parasitized is stupid, it is because it is good to be stupid, because some errors are less costly. I do not think that there would be, in nature, much place for the type of problems such as the arms race. We think, in this framework, that one is stupid and the other intelligent. The strategy of lying is not good unless we assume that others are imbeciles."[19]

To the ontology with Panglossian resonances (all that is in nature is the best possible) will correspond a form of ethical epistemology—that also probably allows the possibility of breaking with the research on invariants: "If I go into nature and I see a strange behavior that I cannot understand, I have two possibilities: either I do like my colleagues at Oxford, and I say that the animal is stupid, or I tell myself that I am the one who is stupid, and I stay in the field until I understand the behavior."[20]

This seems to be the message that Richard Dawkins understood, and which causes all sorts of trouble, when in the second edition of *The Selfish Gene* he addresses a public apology to Zahavi that was supposed to amend the pointed criticisms he addressed to him in the first edition: if Zahavi's hypothesis is correct, which now seems to Dawkins to be the case, the perspective is troubling because it "means that theories of almost limitless craziness can no longer be ruled out on commonsense grounds. If we observe an animal doing something really silly, like standing on its head instead of running away from a lion, it may be doing it in order to show off to a female. It may even be showing off to the lion: 'I am such a high-quality animal you would be wasting your time trying to catch me.' But no matter how crazy I think something is, natural selection may have other ideas."[21]

The saturation of meanings carried out by the anthro-pomorphic, anthropological, adaptationist, and fictional ap-proach is not dissimilar to this characteristic of mythical rationalities, such as they were described by Lévi-Strauss: con-fronted with a real to understand and its absence of meaning, its non-experimentable objects and its experiences without objects, the myth will set out a *plethora of signifiers*[22] and over-determine the real with them.

MYTHICAL SCENE AND EXPERIMENTAL SCENE

To our description of Zahavi's method as a mythical and at-tributive story unfolding between two characters—the human and the bird—we opposed the experimental approach in de-scribing it as a method that triangulates the roles, since it makes three characters take part: a bird-witness, an observer-manipulator, and an imaginary jury. Can we say that this last aspect protects it from all attributive temptation? In other words, are we justified in thinking, according to these descrip-tions, that one of the two approaches is totally attributive and that the other blocks the multiplication of protagonists, the intrusion of subjectivity and the processes at work in mythical rationalities? I recall here two interpretive events: the hypoth-esis concerning the "modesty" of the babblers, first of all, and then that of the "false signal." We will remember that with the hypothesis of social control, Zahavi stipulates that the "mod-est" male who keeps all the other males away from the scene of mating proves its capacity to the female for preventing any future interferences around the nest. We will analyze later the phenomenon of the "mirror" between Zahavi's methods and the way in which the males carry out the eviction of other pos-sible suitors. Consider instead Jon's alternative proposition. According to him the Zahavian reading of the urgency of inti-macy forgets an important character in the scene: the female. The females, too, exert a form of social control, even if in a different manner. They can also enter into conflict with the males, even if this conflict does not take on the appearance of outright conflict. According to Jon, the female can gain this control by hiding from the males the identity of the eggs' father. If the female can hide each act of mating, and if she isolates herself with each male of the group, none of them can be sure who the real father of the brood is. On the contrary, and if the act of mating is really discreet, none of them can be sure that he is *not* the father of the brood. Each of the males has an interest in taking care of the progeny that could be his own.

Before everything else, I open a parenthesis emphasizing that Jon shows one of the important lacunae of the Zahavian system: it is characterized by a rather singular "androcentrism." The male controls the totality of the relations, and what can be interpreted as a strategy of the females becomes an exhibition of the male. We will return to this aspect of the practice theories later on.

Furthermore, we will note that the clear influence of the theoretical framework of the arms race is at work here: a conflict of interests, an evolution of strategies, a parasite—the female— that exploits a host—the male investing his effort in a brood that could very well be another's. The interpretive grid marks, in a very clear manner, the signature underlying the interpretations: rather than an interpretive grid it seems to us to become, in this particular context of emergence of interpretations in "the field," a veritable screen for the production of fictions.

Parallel to the influence of the theoretical corpus, the practices themselves seem to color the interpretation and construction of fictions, as if the observer attributed his own practices to the observed. The females, according to Jon's hypothesis, use means similar to those of the experimenter. Beyond the frame of the arms race, what is described about the female strangely resembles the procedures of experimentation: the creation of decoys—in a sense this is certainly more vague than the usual meaning of the term in ethology—under the guise of the creation of beliefs and the manipulation of subjects. This analogy seems much clearer—and more pertinent—in our second example: the hypothesis of false signals.

The hypothesis "of false signals" regarding the gathering of food for the nest was interpreted by Jon as an "anti-bluff" maneuver. I already attached this hypothesis to the general theoretical model that constitutes the interpretive grid for his hypotheses, and showed that this statement is largely attributed to that frame. We can try to go further and show that here, too, the attributive approach is present in this type of reading of behavior and that it affects the hypotheses, albeit in an implicit manner. What does the bird do, according to Jon's hypothesis? He verifies the real hunger of the brood by setting up a "test" experiment. The experiment acts on a variable to measure the departure in relation to the norm, creates a decoy that allows him to choose between the duplicity of causes, does not put faith in the clues, and makes the nest the site of experimentation. Thus, Jon created a "mirror effect" in his reading of the behavior of the bird by attributing to it an approach identical to the one that he uses. He attributes to the bird his own approach, his frame of thinking, his procedure. Methodology, too, thus becomes a source and model for fictions.

TURNING THE TABLES AND
THE LAUGHTER OF BYSTANDERS

It will not escape the readers that here we are not ourselves outside the scope of mythical rationality when we consider what I call here *mirror processes*. Mirror processes are those that make it such that the contents of the analysis reflect the form or the procedure utilized—for example, when an a priori approach preferentially analyzes the a priori of behavior—or indeed when the researcher seems to consider the animal in the midst of doing something, at the very same time as the researcher himself is doing the same thing: when Zahavi reads behaviors in terms of information, he attributes to the animal the same role and the same work as that taken up by the observer (reading the behaviors of one's companions [congeners] as so many sources of information); in the same way, when

Jon believes that the helpers at the nest test the chicks, he attributes to the bird a type of experimental work in the nest.

This would be the opportunity to stop for a minute to share a laugh with those who have noticed the following fact: I apply this same mythical rationality, and I attribute to each of the researchers the same approach as mine. But those who are laughing, making me the object of identical analysis at this new level of precedence, know that they will also end up creating a nearly infinite mirror game, a regression of attributions that reflect one another, as in the famous Velázquez painting of the mirror that reflects a mirror, ad infinitum.[23] And that they accept, from then on, to laugh with me, and not at me. That is certainly where the lesson of laughter taught by Isabelle Stengers to the ironists resides when she invites us to rediscover together that "capacity to recognize oneself as the product of the history whose construction I try to follow."[24]

Here we come back to one of our very first aporias: the one coming out of the analysis we made of the anthropologist sent to Rosenthal's laboratory, and that showed the impossible exteriority of the researcher, his total inclusion, and the inaccessibility of the fundamental hypothesis.

METHODOLOGIES AND LOYALTIES

If we were to take up the idea of a continuum of methodologies observed in the field, going from the most anthro-morphological (Zahavi) to the most experimental (Jon), a third character appears, whose ambivalence clarifies the practices of the first two from a different vantage point.

Rather than acting as a true mediator, Osztreiher, the doctoral assistant to Zahavi, seemed to me to be the bearer of two loyalties from divergent orders of practice. His practice itself seemed to reveal these tensions: if, on the one hand, he clearly adopts the position of the external spectator, on the other hand, even if he does not whistle at nor feed the birds, his approach proves to be so intimate that it cannot help but influence their behaviors.

Conscious of the demands and necessities of the domain in which he works—which is to say, building spaces of confrontation for fictions—and of the necessity of subscribing to them to be published, he remains nonetheless close to Zahavi's methodology: the management of the space can thus appear as the compromise between the "uninvolved" distance of the objective spectator and the promiscuity of the anthropologist.

I would also study his practice as an attempted reconciliation between the demands of testing fictions in the experimental strategy and the demands of a descriptive and interpretive practice closer to the anthropological approach. If we come back, for a moment, to the disagreement on the subject of the dance of the babblers, it is to recall first that Osztreiher's interpretation remained robustly Zahavian. Then, we could suggest that the dance offered him something other than a site of disagreement: it carries the proof of the influence of the observer on the subjects observed. And through these pieces of proof, by providing him with working material, the dance allows Osztreiher to reconcile the requirements of the two approaches. We note that what he learned from the dance is not totally unrelated to his invasive manner of intruding into groups. His manner is all the more invasive in that he is accompanied in his work by cumbersome video equipment.

The visits to the birds are regularly carried out every day in the morning, from dawn until the end of the morning, and from 3:00 p.m. until the setting of the sun and the settling down of the birds to rest. Some days, the dances can be seen taking place immediately following daybreak or right before sunset.

Recall the two hypotheses at issue: according to Zahavi, the dance, like the ritual, is a test imposed on partners; according to Osztreiher, it is the means of affirming superiority, notably for the division of resources. Osztreiher's research starts from a rather simple observation: the frequency of dances is variable throughout the course of the year. Now, if Zahavi's hypothesis is correct, if the dance serves the function of testing the solidity of group ties, it would need to see an elevated frequency at the time when the group must be the most cohesive, that is to say when the adults must accomplish the most tasks in common. This solidarity is particularly crucial in the period following the hatching of the eggs. Or, as Osztreiher claims, this is not in fact the case. Just after hatching, the frequency of dances diminishes in a rather drastic fashion.

Osztreiher then gave himself the task of comparing the two types of group: on the one hand, the groups that succeeded in the reproduction phase, and on the other hand those that failed during that phase. Once the eggs hatch among the first, he claims that the level of competition strongly diminishes, while this phenomenon is not simultaneously evident in the groups that don't have eggs: the competition remains lively. Among these latter groups, the frequency of the dances remains elevated long enough that there exist possibilities for reproduction. It is only at the end of the reproductive season that we see a progressive disappearance of the dances. A last argument supports the thesis of the importance of "status" in the ritual's function: the adults dance at the time of sexual competition; the young dance more frequently in autumn, at the time of competition around alimentary resources. The access to these resources, during the most difficult period of winter, will be determined by the outcome of the competitions. We could, already, see that the methodologies of Osztreiher seem distinct from those of Zahavi: what, for the master, proceeds from an intuitive development becomes, for the student, the object of a procedure for verification of the fiction. It not only concerns a real confrontation of fictions in a tribunal destined to decide between them—Osztreiher does not just interfere in the natural course of things—but the substitution of one mode of questioning for another: to the "why" of Zahavi, Osztreiher opposes a "how." Zahavi elaborates his theory on the basis of a series clues, and responds to the question of knowing why the birds dance—like a good sleuth, he looks for the motive.

Osztreiher, on the other hand, proceeds to reconstruction: "it cannot have to do with this because of that." Osztreiher looks for proof. If we must continue our metaphor linking the investigation and the trial, I would situate Osztreiher in the intermediate area: the investigation is finished, and now begins the stage of inquiry. We build the supply of witnesses, we study the plausibility of events and motives. With regard to the laboratory approach, we could suggest that the procedure was the object of, with Osztreiher, an initial "purification": from a qualitative analysis, we slide toward a quantitative analysis of these qualities. It no longer has to do, to put it accurately, with a real work of experimentation or the testing of fictions, since it is already within an approach of "watching and doing nothing." However, another part of Osztreiher's research—which is the consequence of his observation work concerning the dances—seems to me to testify to a certain *will to do science*.[25]

Throughout his research, Osztreiher would claim that the frequency of dances not only varied in the course of events of the life of the group, but that it also seemed affected by the manner and the timing of the observer's visits. He thus elaborated a plan for a comparative experimental study, modifying the hours of the visits and the way in which he carried them out (degree of distance, the presence or absence of equipment, et cetera). It was through this modification of his research methods that Osztreiher discovered that visits during the night reduced the occurrences of a morning dance, while morning visits raised them. So he organized an entire work around analyzing the influence of the observer on the behavior of the observed.

Initially, we could surmise that, through this experiment, Osztreiher interpellated Zahavi on the subject of taming practices. But this would not accurately render the manner in which he himself figures the approach of the birds.

If this research reveals that Osztreiher had, in a certain manner, heard the reproaches of his colleagues or had more simply noticed that "something" was going on, it seems to me to indicate an important element supporting the idea that his approach responds to two contradictory requirements. The first requirement, just mentioned, is the one that characterizes Zahavi's practice and consists of observing and doing nothing (not modifying the environment besides any changes that could arise from the fact of the presence of the researcher); the second requirement is that of the experimental approach: act so as to better observe, study the variation in the natural course of things produced by the experimenter. Here, however, it is not the bird that is cited as witness to the process of fictions, but the particular relation between the researcher and his subject–object. In this context, the variables are nothing other than the researcher herself, time, and space; the laboratory is formed in their singular relation. Here, it is no longer the fictions that we put on trial, but the researcher herself who submits to the question. The bird will not be the passive subject of manipulations of its environment, but will be interpellated in a relation that demands its participation. It is in that way that the dance seemed to me to offer him something else than the site of disagreement: through it, he could actively create the conditions for the conciliation of two contradictory demands by which he was confronted. For Osztreiher, this experiment would constitute a way of demonstrating his fidelity to the field practices of his master.

Along with his own new experimental approach in the field, Osztreiher also adopts Zahavi's anthropological method. But in adopting it, he does so in order to extend it all the way to its limit: adopting and adapting with Zahavi's approach, it becomes clear that we are witnessing the result of an anthropological method in ethology whereby he takes into account the fact that those whom he studies will take on the dispositive that studies them and attribute their own meaning to it.

THE IMAGE AS THE SHARING OF FICTIONS

The production of the image itself, through the omnipresence of the camera, can also constitute the object of interpretation in terms of the conciliatory procedure between the demands of the two extremes of our continuum. The production of the image is of the order of pure observational practice, since it does not act on the real (if it is not in the aforementioned frame). At the same time, it fixes the observer in the role of spectator: the dispositive—photographers are quick to point out—quickly creates a distance propitious to the impression of being a stranger to the scene that is unfolding. One does not film in the middle of things, one films in front of someone, in creating an imaginary scene and a background.

The practice of film is no doubt already situated on this same level of pure observational practice, thereby also functioning as a mode of conciliation between two requirements: the activity and the objectifying distance of the experimenter, and the passivity of the collector of anecdotal information. The activity of collecting images seems thus to constitute the mediation between two types of practice. It is also and above all the first step in a process of "scientificization" of discourse, to the degree that it authorizes the sharing in other modes than those of the narrative confrontations of fictions. It allows reproducibility of what is observed. Indeed, if Zahavi bases himself on certain observations to make a hypothesis, no one is obligated to believe it. The solution adopted in general will be either to open the natural laboratory to foreign researchers who can thus verify the statements of the researcher, or that other researchers can try to reproduce or observe in their own experiments or conditions, in order to verify that the phenomenon observed is indeed reproducible. Producing images already offers a first guarantee of what the witness says, it puts discussion material for interpretive fictions at the disposition of members of the tribunal of experimentation. Sharing images is already to admit that the word alone is not enough, and that it will not suffice for the sharing of fictions.

5. Narratives and Metaphors

As we have already had the occasion to see, the anthropological practice of this type of research relies, in part, on the collection of data of an anecdotal type.

The keeping of observational note cards and large calendar charts allows each of the human researchers to know the situation of each of the bird groups: in this way the births, the ejections from the group, the new arrivals, the deaths, and the changes of status are recorded here. Each of the researchers comes to the research center, at the end of the morning or the day, to record the data they have collected and consult those of others there. Zahavi's visits, generally at the end of the week, are the occasions for meetings in which the information collected is discussed.

The laboratory of the center is thus the place of passage and meeting, of formal and informal encounters. Its geographical situation itself favors them since it provides access to the interior courtyard where the lodgings of the different researchers working in the nature reserve are situated (to avoid it by entering through another path requires a considerable detour).

NARRATIVE PROCEDURES

The encounters in the laboratory are sometimes the occasion for exchanges of information that are particular in form: here the babblers become the heroes of narratives that are quite astonishing. The articles that are specifically devoted to them also bear this singular mark. The narrative form can take on the appearance of veritable epics and the researchers swear, with laughter, that they have the impression of witnessing genuine soap operas.

As Stephen Jay Gould observed, "We seem to be caught in the mythical structures of our own cultural sagas, quite unable, even in our basic descriptions, to use any other language than the metaphors of battle and conquest."[1] To the epics of war and conquest described by Gould are added, in the story of the babblers, amorous adventures. The morning conversations around the charts are so many illustrations of this:

> I saw a female. I do not know what she was doing! She cried out to the males, "I am lonely, I am lonely," I don't understand why. Since they were right next to her!

Or in addition:

> A refugee female came to take the place of the female [those we call refugees are the ones who have been ejected by a group, or who have left one, and who have not found new territory], they fought each other. The males did not intervene.
>
> It is true, they are interested in taking the best one. But sometimes they will nonetheless intervene to protect the female.

And sometimes with worry:

> The female (name) disappeared. The other birds are worried about her. They searched all morning.

And in the evening:

> She finally came back. I really wonder where she was.

The articles themselves will adopt, even if in a less empathetic and enthusiastic manner, the narrative and anecdotal genre. This is not only the doing of the anthropological method previously described, but must also in equal measure be related to the very great variety of behaviors among the babblers: for example, the management of sexuality in the group—determining who can couple with whom—can take forms as numerous as the groups themselves.[2] Only anecdotal narration seems able to take account of this.

Incidentally, we find here one of the instances of what we have called "the babbler in discourse": the narrative and anecdotal form is the result of both the anthropological approach and the creative and varied behavior of the birds. It is also here that we are situated in the space of equilibrium that does not allow us to discern—or to decide—whether the birds or the method will affect the other to the extent of making ethology a qualitative approach.

We can tilt to the side of the birds to illustrate the diversity of these behaviors and try, at the same time, not to lose sight of the fact that a particular gaze and sometimes very marginal interpretations were required to give these behaviors not only a meaning, but beyond that the possibility of existing as given, of being recognized as a source of information: "The females can court a dominant very early in the season by giving him a bit of food or a little stick. The male often reacts aggressively. It is then frequent that the female chooses to court and copulate with subordinate males and reject the dominant male. I think that the female behaves in this manner in order to lead the dominant to exhibit his status (that is to say to show his claims in the division of coupling rights). Later in the reproductive cycle, the dominant will tend to remain near the female and she will tend to reject the subordinate males."[3]

But this history cannot be generalized: it happens more frequently than others, that's all. It also happens that "the beta male visits the nest infrequently—or even on rare occasions. When a beta male is allowed to perch next to or almost next to the nest with the alpha male, he will be aggressive with the female beta and will frequently follow the pair of dominants as if the female beta didn't exist. However, if the occasion presents itself during the day, the two males can couple with the female beta."[4]

In the conversations around the data boards, we discover

warlike adventures parallel to the narrations that explain the way that sexual exchanges take place. These relate battles between neighboring groups: the terms borders, refugees, victory, and defeat give these enthusiastic narrations a highly realistic character.

Sometimes the two genres intertwine with one another:

> A female went to the border of the territory. She wanted to provoke the neighbors. She made repeated incursions beyond the border until they responded to her provocations. In fact, she wanted to test the two males of her group. Before laying her eggs and embarking on such an investment, she must verify if the two males are trustworthy and if they are ready—and capable—of defending the territory.

This interpretation, that surprised me right off, is, in fact, a pure product of theory, and I finally found it in an article dedicated to the way that males guard females in the management of conflicts. We have already encountered this hypothesis when we pointed to the babblers' "modesty":

> To be able to succeed in a coupling, the male babbler must be able to avoid the approach of other members of the group. I think that the aptitude of the male in controlling the approach of other individuals shows the female his capacity to control the group in general. This capacity to control the group can, in turn, be correlated with the skillfulness of the male in defending eggs against rival males.[5]

Beyond these particularities of method and of the birds, which we just suggested were responsible for the narrative structure of the discourses, we can argue that this practice of ethology figures the processes of fictionalization inherent in the human sciences in an exemplary manner (and from which most of them try to defend themselves, notably through quantitative methods).

We should recall the moral tales described in the introduction, those beautiful stories that we love to produce and listen to. The theory of evolution itself is a story, and its reconstruction is not immune to the rules of the genre. Science ceaselessly re-entwines with mythology, and the stories told give each of them the mark of the beliefs that participated in its construction.

ANALOGICAL PROCEDURES

Narrative procedures are the product of heterogenous and indissociable characteristics: our mythic mental structures to begin with, followed by the processes of identification or the anthropomorphic approach, then the diversity and originality of the behaviors of the birds, and finally the proximity of the researcher with her object, the individual identification of each of the birds, the almost daily following, the inscription into a story—in sum, the anthropological approach.

For three of these four characteristics, the reference to the human seems to be constant. These three characteristics are generally the most contested in the domain of ethology charged, like all other scientific fields, with assuring and maintaining the great divide.

But the lack of respect for conventions does not stop there. Zahavi's theoretical articles bring up one constant: humans are the analogical reference and the tool for confirming hypotheses. In refusing the process of the litmus test for fictions, Zahavi must find the means of support elsewhere for what he claims. Only subjective experience, shared by all, could fill this role. The reader thus finds himself mobilized by suggestions of identification with the birds, taken to task, and the appeal to his experience will be the argumentative and pedagogical support for the theory. This characteristic renders Zahavian discourse very similar to what Paul Feyerabend calls in his writing on Galileo—and without a pejorative connotation— "propaganda": "It is clear that allegiance to the new ideas will have to be brought about by means other than arguments. It will have to be brought about by irrational means such as propaganda, emotion, ad hoc hypotheses, and appeal to prejudices of all kinds. We need these 'irrational means' in order to uphold what is nothing but a blind faith until we have found the auxiliary sciences, the facts, the arguments that turn the faith into sound 'knowledge.'"[6] "The whole rich reservoir of the everyday experience and of the intuition of his readers is utilized in the argument, but the facts which they are invited to recall are arranged in a new way, approximations are made, known effects are omitted, different conceptual lines are drawn, so that a new kind of experience arises."[7] Zahavi, in his own "irrational" approach of propaganda, makes successive slides between humans and animals, and leads us to identify ourselves with more and more convincing actors. In this he sows the seeds of confusion, scrambles the spaces, and contravenes the rule of the great divide.

But that is not, to my thinking, the only point of the controversy. The effacing of the great division could, in all rigor, still be excused—even if it would be more difficult among us than in the Anglo-Saxon countries. It should suffice, in order to persuade oneself, to take a glance at the ideological and epistemological critiques addressed to the theories of ethology by each side. What would be less excusable, by contrast, is the image attributed to the human in this analogical comparison.

At a colloquium on immunology, Zahavi was invited to discuss the part played by the theory of signals in the processes implicating peptide communications.[8] The international researchers brought together at this conference were, for the most part, biologists specializing in the area of immunology, cognitivists, and doctors. Zahavi's theory was therefore new for them. Zahavi presented them with the main principles of his theory of handicap and reliable signal selection without making reference, in the first part of the exposé, to their application to neurological and biochemical communication. The exposition of his theories thus referred, as Zahavi habitually does, to animal and human examples. The reactions were, most particularly on the part of those researchers coming from France (as well as some American researchers working in a French laboratory), very hostile. Pleasantries were exchanged, as well as gestures of annoyance and reprobation: on one of the notes that was passed, I was able to make out, underlined and punctuated emphatically with exclamation points, *Ideology of competition!!!* One of the attendees of the colloquium later expressed his regret at seeing ethology once again take up such a competitive vision.

The examples in which humans are the reference, it is true, give a none too grandiose vision, and the Zahavian literature gives many illustrations of an anthropology, on the whole, with which we are unfamiliar.

Indeed we witness in the course of Zahavi's writings over time a series of moves that can only confirm his emphasis on competition. In a 1982 article studying the role of vocalizations among birds, Zahavi uses the example given by Sherer that shows "evidence on the correlation between psychological stress and vocal signals in man," and he continues a few paragraphs later: "It is generally known among singers that even small changes in posture and movement affect the quality of their song. Actors on stage are trained to talk in a way appropriate to a particular motivation by imitating the posture and movement of the person possessed with that motivation."[9]

The same example will recur in 1987, this time to take on a metaphorical turn that is clearly more warlike and less innocent:

> The performing actors for whom the task is not to play by the rules with their audience concerning what constitutes their real motivation, know well that to emit a voice that demonstrates a certain motivation, they must adopt the posture that would be taken on by the signaler possessing this performative motivation, and this is possible because there is no risk implied in the performance. The actor can relax in threatening its rival because they don't need to fear an attack. In real life, an inferior that is fighting cannot permit itself the luxury of relaxing its muscles while it is threatening a superior because the risk run from the fact of relaxing will be too great if the rival attacks.[10]

Regarding the necessity of a cost of the signal for the re-liability of the system, Zahavi offers this even more explicitly warlike analogy: "Such an adaptation should imply the un-derstanding of a signal in relation to the circumstances, that is, whether the cost in a particular situation is high or low, whether a consequent evaluation of the particular situation is high or low, and what the signal accordingly means. This we understand intuitively: the man who dares to stand up when the bullets are not flying around is not considered particularly courageous."[11] Our last example complements the former one since it offers an interpretation of threat rituals. To illustrate his theory translating threat displays as handicaps taken on by protagonists to make their signals credible, Zahavi writes that "the cost involved in the danger of exposing an individual to more rivals or predators maintains the level of the threat. . . . Another possible advantage of shouting may be developed from a principle suggested by Schelling. . . . Schelling pointed out that a declaration of intent by a party to negotiations, prior to negotiations, may strengthen the position of the declaring party. The declaration makes it more difficult for the declaring party to withdraw its claims. If it has declared its intentions, but does not achieve them, that party loses both its prestige and the specific goal of negotiations. Hence, after a declara-tion of intent the second party should expect the declaring party to be more persistent in the negotiations. A threat is equivalent to a declaration of intent to use force in order to obtain something from a rival."[12]

In the same way, the definition of ritual undergoes a shift in meaning toward a more competitive function. Where the classical theory assigned it a pacifying role in the inhibition of conflicts, Zahavi's theory focuses attention on the conflicts that can persist in any cooperative relation: ritual admittedly pacifies because it puts the claims of each of the participants to the test. The emphasis is placed resolutely on the competi-tion that is produced by it. Here we reencounter the metaphor that gave rise to our fiction: "It is possible for the observer to people walking in the street to predict from the way they move which of them can hardly run at all and which of them are good runners. But in order to classify people exactly according to their quality as runners, it is necessary that they display their ability to run in a standardized way."[13]

Art itself does not escape from the competitive model since ballet—as we have already indicated—is defined as the very site of this competition:

> A person who does not know classical ballet well will not be able to judge the differences between two good dancers. Even an expert can encounter difficulties in evaluating them, unless the dancers carry out exactly the same movement. I believe that these styles of performance were not developed in order to show that a dancer belongs to a particular style, such as ballet, but in order to develop a standard style of competition between dancers.[14]

It also pertains to demonstrating superiority or inferiority through human vocalizations, since "my experience with human vocalizations, during aggressive encounters, suggests that the same individual threatens with a relaxed vocal signal of a low pitch when confronting an individual which is inferior to him in his fighting potential and raises the pitch of his voice when confronting a superior fighter."[15]

The relations between parents and children do not escape the rule of conflict, and the behaviors traditionally considered as the most affectionate are now defined as so many tests imposed on the partner of the relation:

> Kissing, hugging, and cuddling are communal phenomena. The young of many species can inflict physical stress on their parents by jumping on them, hanging on them, and manipulating certain parts of their bodies (pulling, biting) until it does harm to them. The jump of an infant onto the back of its parent who finally returns to the house after work is a signal of love and affection. Why must a signal involve stress? These observations are understandable when we accept that the relation between parents and children

involves a common interest and a conflict.[16]

Here it seems crystal clear that what I have evoked by the term "propaganda" seems to be at work through the reservoir of the readers' daily experience, albeit with the facts rearranged in a new and different way. The conflict between parents and offspring takes on a much more urgent allure when Zahavi, drawing on Trivers's hypothesis, tries to understand why offspring cry so loudly when they need their parents. The parents find themselves psychologically overwhelmed by their child's cries, and must choose between "two bad branches of an alternative: they can either invest in feeding the child more than they usually would, or risk losing their child from the actions of a predator" (who would be drawn by the cries and could locate the chick, for example). Human children also "force their parents to take care of them by having a tendency to put themselves in situations where, if the parent does not stop them in time, they could fall and injure themselves. An alternative strategy consists in putting themselves in dangerous places like the act of sitting in the middle of the street."[17]

Richard Dawkins called this, not without humor, the screaming baby strategy.

What does this anthropology reveal? Would it be, as social reductionism has sometimes pushed us to think, the result of decades of conflict? Would the zoological theory that derives from it not itself be anything but the projection of our human political rivalries into nature? And the life of the babbler the simple reflection of tension for Israelis between two incompatible ideals: the cooperative and egalitarian ideal of the kibbutz and the liberal ideal of the Western economic model? The fiction of the babbler would then be nothing but the result of an *Umwelt* marked by war, deception, and disillusioned ideals, of a vision of the world marked by renunciation and resignation. When Zahavi speaks of the failure of the kibbutz, he sees in it the proof for the impossibility of the theory of group selection:

In the kibbutz, all the social positions are equal at the beginning. How is it possible to incite people to cooperate in conditions where it is the same thing to work well for the kibbutz or to not do so? A system of cooperation, where the rank is equal for all, can encourage people to do well in order to win or raise their status in the eyes of others. In the first kibbutz, all were equally poor, and the quest for status was undertaken inside the community. But things have changed. Now, people no longer content themselves with status in regard to the local community. They want more, to be recognized elsewhere. The world is larger now. Or else they don't care. Thus, some have stopped working. And the kibbutz has become the site of social parasitism. And the others have to work in their place and feed them nonetheless. This leads to conflicts and dissatisfaction. The kibbutz is the proof that group selection does not work.[18]

We can hypothesize that Zahavi's model strangely resembles an ideal model of the resolution of conflicts between the well-being of the individual and the well-being of the group, a resolution where social status would be the effective compromise. The life of the babblers, in sum, resembles a kibbutz that has not undergone the invasion of modernity. There can be a series of elective affinities woven between a theory of ethology and an economic system in tension between two pressures: that of liberalism and the most strident economic competition, and that of utopia and a state that is caring, welcoming, and protective of the destitute.[19]

But to pretend that the life of the babblers would be nothing but the reflection of the political and social structure of the state of Israel would seem to me, after the path taken to avoid this reductionism, unjust—unjust because this would again amount to saying that the true is explained by nature, the false by society.[20] The space of equilibrium, if it is less incisive in critique, if it lets itself be taken in, if it leaves the maze of ties tangled and some aporias unexplored, encourages maintaining the same respect for the one who speaks and the one of whom they speak.

Inversely, to imagine reducing the vision of the kibbutz to the simple influence of the babblers—as if Kropotkin had been anarchistic because the terrain of Siberia contradicted Malthus—would lead us to revert to the idea that nature itself totally determines our vision of the social.

All the products of this theory would be better off suspended within this space of equilibrium, in the space of hybrids. The status of hybrid between nature and culture preserves the complexity of what initially presided over the constitution and representation of what we would like to see as the weaving together of a beautiful quilt that would unify the state of Israel, economic liberalism, solidarity, and those magic carpet operations: the desert, the soldiers from the war, the kibbutz, the babblers, the dances and the gifts, the bread distributed, the researchers, the handicap theory, and so forth.

This status of hybrid can perhaps save us from the anthropocentrism that shadows every step of my path, since by dint of wanting to make an anthropological lesson of zoology I am myself ceaselessly tempted to efface the pole of nature from the domain studied.

Would I not then be in the process of forgetting that I, too, had seen the babblers dancing?

6. Models and Fictions

I t is remarkable," Karl Marx wrote to Friedrich Engels in 1862, "to see how Darwin rediscovers, among the beasts and plants, the society of England with its division of labour, competition, opening up of new markets, 'inventions' and Malthusian 'struggle for existence.'"[1] Engels would take up and complete this critique, in a letter to Lavrov, in 1875, in words that seem to concern social Darwinism rather than the theory of evolution: "This feat having been accomplished (that is the transposition of society to nature), the same theories are next transferred back again from organic nature to history and their validity as eternal laws of human society declared to have been proved."[2]

This sleight of hand is what Richard Lewontin called, in his critique of sociobiology, the process of secondary derivation.[3] As Marx notes, this process in its initial stage often takes on the allure of the naïve application of our social categories to the natural world.

SECONDARY DERIVATION AND BOURGEOIS STRATEGY

In the introduction I underlined the fact that the bourgeois strategy would constitute an example of anthropomorphism. As it happens, the use of the concepts bourgeois and property seems to bring a natural legitimacy to our economic system founded on property, since the evolutionary game that describes the bourgeois strategy considers property to be the regulatory mechanism for conflicts. To situate this, recall the hawk–dove strategies. The evolutionarily stable strategy, as we noted, is composed of an unstable equilibrium between two strategies. The model becomes more nuanced and stable with the bourgeois strategy, which consists of adopting the hawk strategy when one owns a good coveted by others, and the dove strategy when one is the coveter. The fiction of the artificial laboratory shows that the bourgeois strategy brings, on average, the biggest gains.

The examples, taken from sticklebacks as well as from the baboons filmed by Hans Kummer, seem to confirm the validity of the model. But in reality all that we can say in observing sticklebacks is that an animal who participates in a territorial dispute shows, on that occasion, a conflict of motivations or contradictory emotions: the motivation to attack or the aggressive emotion, the motivation to flee or the emotion of fear. Now, what we can note is that the more a stickleback is at the center of its own territory, the stronger its motivation to aggress seems to be, and the further it goes from its territories, the more the motivation to flee seems to augment. This is what the bourgeois strategy describes, and nothing else. What the strategy does not permit us to say—but which is validated nonetheless, as the phrase of Maynard-Smith shows—is that property is considered, by convention, "as the decisive parameter that regulates conflicts."[4]

One could say that for the process of secondary derivation, a metaphor is somewhat emptied of its metaphorical, analogical power. The metaphor thus becomes a dead metaphor, a catachresis, and there is no longer another term to designate what it refers to. When metaphors are transformed into catachresis, models acquire the power to legitimize secondary derivation and to naturalize moral, economic, and social systems. Darwin must have intuitively predicted this when he put readers of *The Origin of the Species* on guard against the direct and non-metaphorical use of the notion of *struggle for survival*. We can see what he warned about with Herbert Spencer and the eugenicists who followed.

In Zahavi's theory two metaphors capture our attention: that of hierarchy, first off, and then that of cost.

The concept of hierarchy is, for some years, the target of many criticisms. According to Lewontin, the concept participates in the process of derivation: as the concept is reified, it induces the belief that hierarchy belongs to natural categories. Very often, partial observations lead to summary categorization. One can remark, for example, that what a male ethologist describes as "dominance" is much different than what a female ethologist means by this term: among the former, dominance refers to helping oneself first, shoving, threatening, taking. Among the latter, "the word relates to a manner of weaving around oneself an affective network, a series of affiliative exchanges, of agreeable communications that structure the group," writes Boris Cyrulnik in commenting on the work of Shirley Strum.[5]

The studies on what has been called the hierarchy of dominance generally tend to focus on one sole parameter, for example the access to food or to females. There are, however, good reasons to think that in many species the position occupied by an individual in a hierarchy of dominance, based on a given parameter, will not coincide to the position occupied by the same individual in a hierarchy based on another parameter.[6] We can credit Zahavi with taking account of this second parameter and of adjusting and refining the categorizations, in sum of being able to—and for the first time—notice something different in what had seemed similar. But nothing would permit us to think that only these two parameters need to be taken into account. In effect, this testifies to a remarkable characteristic of the theory: as we have emphasized, following Jon, this story singularly lacks an actor. The female does not show up here except as in the pale role of the extra. Dominance and the social control of the group are always considered from the male's point of view. The power or status of the female, itself, is never considered. Who posed the question of knowing whether the females dance? The theoretical system is entirely built on the strategies of half the species, never taking into account the other half. This is not without some ties with the context of justification itself: in the field, whatever the status of the females, only male ethologists carry out real research in Zahavi's territory. A young, female, American postdoc—whose qualifications were, it would seem, of a very high level—had been sent to the field to observe the babblers. This researcher worked in a laboratory in the United States carrying out very specialized research on the birds' genetic code. The lab director she worked for sent her to Israel so that she would have an opportunity to "see what there was fleshing out the DNA." Despite her high level of competence, the young researcher was assigned all the subaltern tasks and technical inventory. It would seem here that there is, clearly, an "elective affinity" between a very asymmetric system of cultural organization of the relations of power and ornithological theories that struggle to recognize a true social role for females.

The metaphor of cost poses its own share of problems: nothing, in effect, allows us to claim that the links established between the costs and the statuses, in the human examples chosen, would not themselves be the result of a process of secondary derivation, of the metaphor's change into catachresis:

> The process by which a signal can be lost due to a reduction of cost is analogous to the inflationary process. . . . Even if the cultural system were not a simple evolutionary–biological process, it would seem probable that it would obey similar rules to the degree that it is also a system of competition. From the 16th to the 18th century, lace was largely used in Europe. At that time it was made by hand and cost more than gold. The price of the lace was steeply reduced when weaving machines were introduced in the 19th century, and this reduction allowed most people to wear lace to decorate their clothes. While formerly the people who used lace to decorate their clothes used a great quantity, its common usage ceased in a very brief period after the mechanization of production. I suggest that this lace went out of fashion because it could no longer be a reliable external signal of wealth.[7]

And some years later, following the same line of thought, Zahavi wrote, "The handicap principle creates a logical connection between the detailed pattern of a signal and the message encoded in it. It suggests that for every message there is an optimal signal, that best amplifies the asymmetry between an honest signaller and a cheater. For example, wasting money is a reliable signal of wealth because a cheater, a poor individual pretending to be rich, does not have money to throw away."[8]

This classification in terms of cost originally emerged from the ways of thinking about the game theory of evolution, but in identifying cost and reliability, however, it seems to have gone beyond the frame. The transformation of the metaphoric cost (expenditures of energy, time, resources, risks in the face of predators, et cetera) into a catachrestic cost (the money we spend) points us toward this slippage that is being carried out: the reliable signal seems to be a concept close to that of credibility for Pierre Bourdieu. If we refer to the analysis Bruno Latour makes of this, we could think that the new classification and its metaphors facilitate the synthesis of economic and epistemological notions (certainty, doubt, proof): the credit given to a signal is both an economic credit and an epistemological credit, "what furnishes the observer with a homogenous vision and scrambles the arbitrary divisions between economic, epistemological, and sociological factors."[9] This confusion, this jamming, no doubt masks a certain number of issues, and the love story between biology and economy is a veritable nest. But the denunciation of this confusion is itself revealed to be a site of concern, and it shows us one of the signs of our intense attachment to the separation of domains.

SOCIAL MODELS, CATACHRESIS, AND *UMWELT*

The process of derivation at the origin of this confusion arose from a two-step process. In the first step an anthropological model of a particular type is applied to understand or describe an animal: the social model. Indeed, it is thanks to this model that we "recognize" a model of their own society among animals—different from the simple identification model that allows speaking of animals as seeking, thinking, or dancing. While in the second step, we see how a sort of "naturalization" of the human practices that played the role of model in the first step of categorizing the natural world is carried out. This double operation evidently takes place in an implicit manner in a procedure that effaces the traces of the passage from one pole to the other, and that removes the history of the conditions of its production. Thus, in a shocking sophism, the bourgeoisified animal not only legitimates, but even, in a pinch, moralizes property as a natural right that pacifies the world by regulating conflicts. It is this process that transforms, in my eyes, a metaphor into a catachresis: the bourgeois or the cost no longer designate an analogy, in the same way that the crowbar no longer designates the object by analogy with its natural form, but designates the thing itself.[10] The proof is that there are no other words to refer to it.

It is here that the limits of Zahavi's theory appear. Of course, when the metaphor remains alive it does not become a process of secondary derivation, but an analogy with a heuristic or pedagogical value. Only attachment to the great divide can make critique of the procedure possible. But when the metaphor is no longer one, the process ceases to respect the living logic of comparisons, and founds a sophism the classic—and amusing—form of which generally presents itself in this manner: all humans are dust, there is dust under my bed, therefore there is a man under my bed.

The process of derivation was not the object of critique for any of the commentators on Zahavi's theory. It is possible that the fact of being immersed in a less-scientific Latin European culture made me more attentive—and more susceptible—inasmuch as we blend the human into the matter. It is also possible that the habits of ethologists of recent years, and above all their way of expressing their theories (think, for example, of Richard Dawkins), have created a metaphorical field where the conventions are so widely known and implicit that there is no need to recount them. The examples drawn from humans are perhaps nothing, in this perspective, but an almost conventional manner of being pedagogical and creating a realistic effect.

If this critique was not, to my knowledge, explicitly formulated, others have not failed to call the validity of the model into question.

Empirical limits were formulated, either because the experimental practices did not confirm the theory (recall here by way of example the experiments of Slotow on sparrows in which the color of their plumage changed), or because it contradicts the logic of evolution: for example, Marian Dawkins and Tom Guilford point out that the model does not give enough importance to the cost run in the act of receiving the message. A message can also be costly for its recipient, in terms of loss of time, the risk of being infected by diseases that the signaler might carry by staying so long in proximity to him, to which must be added the fact of being more easily attacked by a predator in remaining next to a conspicuous individual, and on and on. As a consequence of this, the more costly the message is to the signaler, the more it can equally be so for the receiver. This would have to have the consequence, from the point of view of evolution, of diminishing the costs for the two parties by substituting for costly signals other signals that are more conventional, but clearly less reliable.

Other elements coming from the game theory of evolution would also seem to call the theory into question. We will review them briefly.

Another limit appeared from the pen of Marian Dawkins, questioning the necessity of the cost in the signal's definition. Defining the signal as costly inscribes the reasoning in a recursive definition: What is a signal? It is costly information. What do we call costly information? A signal. But this definition, in its rigorous simplicity, actually constructs an artifact.

In fact, isn't a signal a signal that appears to be so to us? And our *Umwelt* is such that only signals costly to the animal are accessible to us. A conspiracy of whispers (as Dawkins and Guilford put it) within a relationship with a low level of conflict, such as that found among dolphins, will not be defined as an exchange of signals, and will not be taken into account in the analysis of its parameters and conditions.

The world delimited by our *Umwelt* primarily opens us to signals that resemble the signals we use with each other. And if the signals strain our sight and hearing, it is because the result is a considerable expense. The signal is therefore not costly except in that our visual and auditory system lets the signals which would be below that threshold escape. Thus the female of the black grouse chooses her male according to particular criteria to which we do not have access. However, the choice is not made randomly, and Rauno Alatalo, Jacob Högland, and Arne Lundberg had good reasons for believing that criteria exist: all the males chosen by the females were still alive six months after mating without our being able to know what allowed the females to predict this so accurately. The signal defined in terms of cost would be, according to this perspective, but one of the subsets of the category of signals.

Marian Dawkins's argument revisits, in some sense, the classical form of the controversies found in ethology—of which we have already spoken—since the critique calls into question the ability of the observer to see what must be seen or to create what must be created. The argument here is not against one observer in particular, but rather against the generalization of theories too dependent on our *Umwelt*.

CONTEXTS OF EMERGENCE AND OBSTACLES

We could think that a great deal of time was required for Zahavi's theory to be accepted. If the expression is justified in its rhetorical sense, it was not so in its literal sense. Time, in fact, does nothing, and it was a fact that the theory emerged at a decisive moment. But it is also true that Zahavi had to wait a long time in indifference and incomprehension. And this indifference and incomprehension were not unrelated to the fact that he "does not play the game," which is to say that he does not experiment, thus cannot pretend to impose his fiction on others, and that his anthropomorphic and anecdotal language does a disservice to him in scientific journals. The "scientific game" is not only ignored by Zahavi, but he can even come to overtly mock it: "The babblers are marvelous birds for ethologists. When we make an experiment, we make predictions. If the birds do not do what we predicted, we teach them to, and then we redo the experiment and they confirm what we predicted."[11]

Everyone—perhaps amid the effectiveness of their *propaganda*—thus had to oppose the acceptance as much as the rejection of the theory. I insist here on the fact that the theory provoked indifference and not a phenomenon of rejection. A theory that *interests* no one—as Isabelle Stengers has underlined—gives rise to no controversy, and falls into oblivion without being sanctioned by those who "play the game."

The same factors that favored the emergence of an original theory played against its reception in the scientific milieu: we can impute to the distance of Israel and the University of Tel Aviv from the large university centers the possibility of creating hypotheses and methods that are so outside the academic mainstream—elsewhere, Jon told me, at Oxford or Cambridge, they would undoubtedly not have been able to be produced, and the pressures would have been different.[12] But what played in favor of the theory thus played against it as well: the model remained, in the eyes of many ethologists, a marginal and relatively fantastical approach. The anecdotal character of the facts and the refusal of experimentation did not allow the theory to fulfill the necessary criteria in this area. At the beginning of the 1980s, Zahavi's ideas themselves seemed to constitute an obstacle: a scientific journal refused some articles based on the fact that what interested him was altruism and not pseudo-altruism. To these elements that seem to make up a good illustration of the context of justification should be added the more technical elements of a new type of laboratory. These could, effectively, demand the testing of the statements involved in the handicap theory.

PROBABILISTIC MODELS AND
THE PERFORMANCE OF CHARACTERS

Mark Kirkpatrick, with his provocative article title announcing "The Handicap Mechanism of Sexual Selection Does Not Work," shows that the probabilistic genetic models argue in favor of the classical hypothesis.[13] This, he reminds us, stipulates that sexual selection not only can be un-adaptive, but beyond that, can run contrary to natural selection.

The deployment of this demonstration merits some consideration. It pertains, ultimately, to the presentation and validation of two antagonistic hypotheses: the first, classical hypothesis of R. Fisher, postulates that the preference of females will act on the evolution of males, in a manner such that sexual selection will act against natural selection and produce forms that negatively affect their chances of survival; the second, recent theory postulates for its part that the preferences of females and the force of sexual selection generated by these preferences will develop in an adaptive manner and increase individual selection. The confrontation of fictions is not carried out here in the form of an investigation nor a trial. Rather, it resembles the "bringing to reality" of entities "purified" in imaginary tests. Here, the fictional laboratory resembles that of the chemist, or of the alchemist who is the creator of future lives.

The fictional objects deployed by Kirkpatrick are, in effect, particular objects: they are the loci and their alleles. In a way, the objects play the role of masks in the Italian theater: they determine the characters. The characters are those of viability v, of handicap h, and of the preference of females for the handicap p. These characters will enter into interaction in the tests that are the equations, and create systems of elective affinities between them. These affinities are expressed in the mathematical terms of average frequency (their distribution in the population), differential frequency (the effect of differential selection), as well as covariance and correlation (how the frequency of one diminishes or augments in relation to the frequency of the other). Each of these premises is introduced into a system where they will be reiterated a certain number of times, up to the point where the system attains equilibrium. Each of these iterations constitutes and defines the phenotypical frequencies typical of each generation. They thus create so many imaginary future lives starting from simple elements. I won't go into the technical and fastidious details of this model.

It is nevertheless interesting to note that the stakes of this simulation, starting from the basic premises, appear rather clearly: the handicap mechanism is entirely indeterminate from the point of view of evolution, that is to say for a given set of parameters very different results can be arrived at when equilibrium is attained. Female preferences can therefore stabilize around any arbitrary degree of development. Kirkpatrick, with his demonstration, defends the idea of total indeterminism of secondary sexual characteristics; and then, the fact that this coevolution can turn out to be strongly against adaptation and that it is at most—or best—the realization of a compromise between natural selection and sexual selection. These characteristics are generally those of the theorists critical of pan-selectionism. Their presence in this article allows us to better understand the source of discord with Zahavi's theory that is, in many regards, hyper-adaptationist.

The simulation of fictions cannot thus sum up the trials of all equally possible narrations. Each among them receives, on its introduction into the computer in the form of an equation, a task to resolve that is not necessarily described in the fiction—but can sometimes be in their context. The representation through which Kirkpatrick defends his theoretical moves seems to omit some characters from the play, and perhaps even characters altogether. In reducing the role of the handicap only to a signal toward females, and its evolution only to the interaction of preferential choice and the burden it constitutes for survival, Kirkpatrick carries out a double reduction. The first consists in envisioning a world of simple genotypes (the actor is the mask?), a reduction of the level of the living to the "chemical" level of simple entities. The second, tied to the first, reduces the function of the signal. We have had the occasion to see that the signal may have been selected thanks to multiple associated functions: so, a signal that indicates the value of a male to the female can, at the same time, inform the predator of the difficulty of catching its prey; the example of the leap among antelopes or gazelles is but one of many possible examples of this. The viability of the male will thus not be affected in a univocal manner by the presence of a handicap, and the interactions of the three characteristics (v, h, and p) must then become singularly more complicated. One could continue in imagining the effect of the handicap on the one who bears it: female preference alone will no longer play a determining role, notably in the facilitation of the bearer's access to resources. The survival of the female herself can be taken into account since the choice of an extravagant, and thus conspicuous, male must certainly affect her own viability in one way or another.

The source of disagreement with Zahavi's theory is double: first, at the level of epistemological and ontological choices (indeterminism and the arbitrariness of choices, the non-adaptation of selected traits); then, at the level of the representation of the model inasmuch as the genotypes built by Kirkpatrick lead to quasi-nomadic phenotypes where the correlations are limited to isolated and artificial factors.[14]

Mark Kirkpatrick goes back to the mathematical models of Maynard-Smith, Davis, and O'Donald both to modify them and to confirm their validity. These models all seemed to compromise the explanation for the handicap mechanism. It must be pointed out, however, that the mathematical models of Eshel, Wrest-Eberhard, Dominey, Nur and Hasson, as well as Kodric-Brown and Brown, built between 1978 and 1984, arrived at the opposite conclusion that the handicap theory could be artificially validated.

According to Zahavi, and despite this controversy, it seemed that the theory had to encounter several more years of difficulty before being seriously considered. We can inquire as to this seeming indifference by bearing in mind that journals still published articles during this time—and that big names of sociobiology weighed in on the model, to contradict it, perhaps, but nonetheless devoting to it time, energy, and publications. Nevertheless the feeling of non-recognition persisted for Zahavi. As a result he asked that they invite him to Oxford to explain himself because, he said, up to that point no one had taken the trouble to understand the theory's totality. This was done, but he came back with the impression of not having been heard. More generally, the researchers, according to him, had not understood the proposed model.

We can propose a plausible hypothesis here: Zahavi, once again, does not "play the scientific game." Indeed, it seems that when we review the literature published on this subject in the course of the years between 1980 and 1995, he never responds to the critiques addressed to him in the articles. Only Heinz Richner's 1993 article would receive a published commentary.[15] In addition, one should note that this article did not call the theory into question, but proposed an enlargement of the definition of signal—to include elements of the inanimate world such as the size of the territory since it could play the role of handicap—and some clarifications on the notion of ritual.

Zahavi, in not playing the game of a public and conflictual exchange of fictions by way of articles in the form of responses, compromises what constitutes the very essence of scientific work.

If the article critical of sociobiology that he published in 1981 seems to respond, in a rather general manner, to the critiques addressed to him regarding the mathematical models, it nevertheless contains only one response that would allow the conflictual exchange of fictions: it situates itself on another terrain, and contests the artificial site of the experiment. According to Zahavi, mathematical models can be of no utility for or against the theory because they were applied

> to interpret social interactions using simple probabilistic genetic models (e.g., whether it is a better strategy to attack or to flee). Actually, animals I have known in the field react in a highly variable way. The same individual will attack, threaten, flee, or avoid interaction altogether, according to differing circumstances. Reaction seems to be determined by information gathered rather than by a pre-set program activated by simple arbitrary signals.[16]

THE MIRROR GAME IN THE EXCHANGE OF FICTIONS

If one recalls Zahavi's hypothesis concerning the way a male "deters" subordinate males when they try to interfere at the moment of mating, one can find, once again, the "mirror effect" set out when we established our hypothesis. Now that we have analyzed the way Zahavi "manages" the scientific game, we can advance another hypothesis. Observing his strategy closely, what is he doing when he does not "play the game"? He refuses direct conflict by always situating himself on another terrain than the one where the debate is situated. In more vivid terms we could say that when others interfere in his relation with scientific objects—by proposing alternative fictions to him following the conflictual models of contradictory debate or real or virtual experimentation—Zahavi tells them stories, ignores them, or affirms his theory without giving "reasons," without entering into the conflict—even implying that if they are not in agreement, it is because they have not understood. This singularly resembles an argument over authority: not wanting to give his "reasons" is to simultaneously refuse the scientific game, the exchange founded on rationality (reason), and, at the same time, the game of the democratic debate of representation at the heart of the tribunal that rules on conflicts and fictions. Rather than an analysis of the meanings of this strategy, I would propose a surprising analogy that forms this "mirror game": Zahavi uses the handicap strategy he saw at work among the alpha males he observed when a subordinate male interfered in procreation. He refuses the conflict and affirms his superiority and status by situating the response on another terrain, another level: the bird, we recall, cleans, offers a gift, or more to the point, ignores the rival. It is a handicap to the extent that it even affirms by this performative exhibition that "I have the means not to enter into conflict, I can withstand this intrusion, I have the means to move from this conflictual method where the other would engage me." Zahavi, in refusing the game of conflict and scientific debate, perhaps affirms, in the same manner as the bird, that the theory possesses in itself the power of evidence, the power to prevail without the need to pass through the tribunal of experiment.

MATHEMATICAL MODELS AND
FICTIONAL LABORATORIES

An important character occupies the front of the stage throughout this story: Richard Dawkins. We have already had occasion to make reference to him, but without mentioning that he plays a very influential role in the domain of ethology: his books have found great success in both university and mainstream settings, and he has collaborated with Marian Dawkins, John Krebs, Helena Cronin, John Maynard Smith, along with many others.

In the first edition of *The Selfish Gene* (1976), Dawkins voices bits of commentary that are skeptical, even mocking: the handicap principle is "shockingly contradictory" and "gets stuck in the throat."[17] He then invokes logical arguments: "I proposed that the logical conclusion would have to be the evolution of males with only one leg and one eye. Zahavi, who comes from Israel, instantly responded to me: 'Some of our best generals have only one eye.'" The logical argument—

that like many of Dawkins's arguments seems more harmful because of the joking tone and caricatural examples than the logic of the argument itself—follows the argument from authority:

> As long as it is formulated in words, we cannot be sure if it works or not. Up until now the mathematical geneticists who have tried to capture the handicap principle in a working model have failed. Perhaps this is due to the fact that the principle cannot work, or that the mathematicians were not smart enough. One of them was Maynard-Smith, so I would favor the first possibility.

Some years later, in 1989 to be exact, in the second edition of the same book, long passages do justice to the handicap theory. Studying the changes between the two editions shows us where the event played the role of lever: to the passage that we just read, an important note is now added, of which this is an extract:

> Zahavi's theory now seems much more plausible than when I wrote that passage. Several respectable theorists have recently started to take it seriously. What worries me the most is that this list includes Alan Grafen, of whom it has already been said and written that he has the annoying habit of always being right. He translated Zahavi's ideas into a mathematical model and showed that it works. And this is not an esoteric pastiche of Zahavi's ideas, but a mathematical translation model of Zahavi's ideas themselves.[18]

I had referred to an argument from authority: this one seems to me to fit it to the letter, as we have seen it defined by Bateson as the power to bring others to align reality with your theory—and for that matter this could tell us a great deal about the game theory of evolution. Alan Grafen, a former student of Richard Dawkins, put his teacher in a difficult position. In effect, at the beginning of the second part of his most important article, he puts the two theories that had tried to apply adaptationist principles to animal communication side by side: those of Richard Dawkins and John Krebs, who insisted on the fact that the animal cheats and manipulates the other when it communicates, and that of Zahavi, who believes that only honest signals are selected for. What Grafen calls "resolving the conflict between two theoretical traditions" comes down, in fact, to showing the plausibility of the handicap principle as an evolutionarily stable strategy mathematically, and thus as given in a certain manner that Dawkins and Krebs are explicitly wrong, since the implicit definition of an ESS requires that once it is installed, the strategy no longer allows the invasion of a mutant strategy into a system of signals.[19]

But the resolution of the conflict does not stop there, and Grafen will propose some distinctions that do not exclude the possibility of manipulation. These manipulations exist and can, in very precise conditions (a minority of manipulators, for example, or the fact that most of the time cheating brings no advantage), be stable from an evolutionary point of view. Grafen shows that these manipulations—whose existence he therefore recognizes—must be distinguished from the signals of a system of signaling. At the heart of such a system, the two partners both draw—at least on average—benefits from the exchange of information. On the other hand, in the manipulative system, some signals can be exchanged, but this does not constitute a signaling system since only one of the two partners profits from the relationship at the expense of the other. Recall here that the signal is distinguished from other information in that it was selected in the course of evolution. This implies that if one of the two partners of the signaling system turns out to be systematically the "victim" of the signal, then its chances of survival will be lessened, and by the same token its reproductive success and therefore its chances of transmitting its "naïveté" to its descendants. For a signal to be selected, it must be, consequently, that the recipient benefits from the signal. A dishonest signal, from this point of view, will not be selected.

Grafen's model shows that there clearly exists an evolutionarily stable pair—as, for instance, the hawk–dove pair that we have already considered—that fulfills all the conditions of the model in a biologically plausible context: "honest" males (handicapped by their burdens in such a way that the bluff is impossible) and confident females who choose them and advantage their own reproductive success by the choice.

An external critique that analyzed this history in terms of the balance of power (Grafen works at Oxford, along with Dawkins and several other big names in contemporary ethology) and above all in terms of authority and recognition, would no doubt allow the understanding of ties that go beyond the frame of strictly epistemological considerations.

We could also analyze the metamorphoses of "nomadic concepts"[20] that, in Grafen, travel between economy and biology, not this time as borrowings of concepts or methodologies of calculation but in terms of pure application. The handicap principle becomes the nomadic concept that explains the offer of products by means of advertising, the major difference being that only, in principle, the consumer—the receiver of the signal—pays the price. I leave to those more competent than I the analysis of this enterprise of external critique.

This mathematical model finally leads us to a surprising paradox: Zahavi's theory was validated thanks to models that he always refused to use—and that he explicitly called into question, on the grounds that they are both simplifying[21] and simultaneously uselessly complicate things, since they put into equations things that logic alone leads us to accept.[22] This logic claimed by Zahavi does not seem to be shared by all, and Dawkins's remark regarding the craziest ideas seems to resonate with Grafen when he proposes reconsidering certain explanations rather than rejecting them on the pretext that they are absurd.

Here no doubt resides the essential quality of the fictional models, and the considerable contribution that they gave to the handicap theory. The computer here becomes like a laboratory, a fictional space of encounter for fictions, or "narrative matrix" as Isabelle Stengers puts it, that is capable of giving rise to even the most absurd stories and combining elements to prove their validity.

To the question of knowing why Grafen's model succeeded where others had failed, the following hypotheses can be proposed: other authors had initially postulated that animals in good health used a different strategy than those who were not, that they therefore developed different systems of publicity. Now, this is in contradiction with Zahavi's model, for which each demonstrates its real value. In other words, the fictions encountered were not all the same as they were claimed to be.

Following this, authors generally did not understand one of the strategies proposed by Zahavi, acting instead as analysts who took only one extract from an author to understand the meaning of his whole oeuvre. Now, Zahavi—at different times and in different articles (and the credit belongs to Grafen for having been attentive to this)—proposes several different approaches for the handicap principle: the qualifying handicap, by which the male who has survived with the handicap indicates to females his quality in other areas; the revealing handicap, by which males exert themselves to impress females and show them qualities that are not immediately apparent—like the courage of the raven or the babbler, or yet again the energy in the bellow of the deer; the conditional handicap, by which only the males of high quality can develop a handicap; and, finally, the handicap of strategic choice—males have information on their own value, they do not share this information, but use it to "decide" to develop this handicap or not. Grafen was not only attentive to rendering all the possibilities set out by Zahavi faithfully, but shows his talents as a scenarist in reproducing a multiplicity of sufficiently disparate elements while also giving a coherence and a veritable "consistency" to the story by making all the consequences explicit.

By permitting a proliferation of stories and offering a quasi-public space of discussion and evaluation to a fiction that had its origin and evidence in a forgotten corner of the desert, the artificial laboratory allows the Zahavian fictions to be something other than some marginal theories based in conditions that are just as marginal. The theory cannot help but surprise, and I think that it must have surprised more than one among us.

Does Richard Dawkins not write, in his revised and corrected edition, that the principle must have aroused, among Darwinians, "grave doubts"? But he also asks of them "to leave aside their prejudices . . . , [because] the Zahavi–Grafen theory, if it is true, will turn topsy-turvy biologists' ideas of relations between rivals of the same sex, between parents and offspring, between enemies of different species."[23]

It is thus to the future that we leave the last word and the power to verify Dawkins's fictions. It is to the future that we entrust the not yet untangled strings of our path and its aporias.

We will now reconsider, for the last pages, what remains of some of the loose ends of the knot we have taken up.

Conclusions

e arrive here at the moment of revisiting our untidy web
and untangling it to look at the motifs that have been
untied. I would no doubt have been a pitiable Penelope: few
things allow us to see a pure thread, and in the end our web
shows us more reasons for our incapacity to untangle the motifs than threads untied.

In sum, each aporia shows us the limits of critique. But at
the same time it has given us the direction to take to continue
the study.

Isabelle Stengers gives us the gift of two tools that add heart
and courage: the choice of laughter rather than irony, and the
idea of the experiment as place of exchange—exchange at the
heart of the experimental relation that makes of it an experiment productive of existence, exchange as a public place of
confrontations and putting to the test of the ability to produce
fictions.

The laugh thus accompanied me starting from the narration of Rosenthal's experiment, and showed me that we are
henceforth caught up and tangled in the ties that we try to
untangle.

If we wish to learn from ourselves in studying the discourse on the animal, we must first learn to laugh at our in-
competence in the face of an always opaque relation in which
we ourselves are enmeshed. The laugh will follow us to the
moment when we find ourselves so much tangled up in these
ties that we will find ourselves—and everyone this time—with
the tables turned.

The contexts of justification themselves reflect this complexity of ties between theories, practices, and the discourses that take account of them. So the theories that form the context of justification for the handicap theory will themselves encounter a context of justification. Ideological critiques describe it wonderfully. But these critiques cannot understand themselves without their own affective and ideological context. Their contingency leads us from then on, little by little, to a regressive analysis. As in the analysis of Rosenthal's experiment, regression puts us in the position of ourselves being one of the strings of the knot that we are trying to untie. And to the mythical rationality of the discourses by which we tell our stories and that structure them according to the stages directing our moral tales, epistemological habits, and our ontologies, responds the rationality of our analysis, just as much permeated by the same mythical operators. And our analysis, as in the stories of pursuit where the hunter comes to resemble the prey, itself also bears the mark of the myths that accompany and feed all research.

We must therefore exit from armchair analysis and notice the fundamental oblivion: the principal actor of the story—the babbler—was absent from it. Indeed, no controversy without contested object, no representation without object represented.

From here on the analysis of the theory of signals must look into not only the scientific, emotional, mythic, moral, and sociopolitical context, the methodologies and modes of testing fictions, but above all at the babbler, the bits of bread that they accept, the incomprehensible conflicts, the gifts, and the dances.

But would these theories that describe the babbler have been able to emerge if the gaze of the observer had been posed on them differently? If the dance was nothing but the exchange of conventional and specific gestures rather than the signature of singularity, the place of testing of the other, if the gift had been nothing but a means of sharing, if behind the similarities differences had not emerged?

And the quality of the gaze, could it have been so if the hand had not been able to take the bird and be cleaned and attacked by it, if the bird had not been zoomorphic, if it had not accepted the sharing of bread? Would the bird have been able to be what it is if the physical space had not been structured like an anthropological space, and the mental space like a continuous, attributive, and anthropomorphic space; if the time had not become the time of the story but rather the variable and manipulable time of experimentation?

Here we find neither a fortunate coincidence nor a conspiracy of events. Here our third aporia appears: the motif can be untangled, but it can reveal to us neither its meaning nor the significance of the knot. Each of the poles remains whole, each one irreducible to the other. Mirroring processes appear: the approach affects not only the data observed, it itself becomes content of analysis and affects the hypotheses. From then on certain fictions will be the reflection of what the researcher was doing at the time of observation: with Jon, the bird tests a hypothesis; with Zahavi, it becomes a competent, solicitous observer of differences.

Going further, research and approach intertwine: the babblers become active in the questions that are posed to them, and express to Osztreiher some critiques concerning his methodology.

The dance of the babbler constitutes, in this regard, an interesting knot. It gives its flavor to the motif outlined. The choreographic space includes, then, the researchers and the warrior epic fiction introduced, the gifts following conflict, the screaming baby and the experimenter, the history of lace, the investigation and the trial, cameras and images, communicative cooperation between predator and prey.

The babbler in discourse is a hybrid object, held in the threads of a web made of beliefs, mentalized social structures, built also from an Israeli society with its kibbutzim and its egalitarian utopia and the economic difficulties that prevent its realization, with its interminable war and kilometers of border. But it is also a bird that actively participates in the discourse produced on its subject.

The quest of the actor in this ethological comedy is thus not ridiculous. What I looked for, without naming it at first, that space of equilibrium that emerges at the end of the strings, took the form of a web of relations in which each was enmeshed. Narrating the handicap theory, in finding threads in this tight web, was thus to narrate a web of funny, moving relations that led me to imagine that the real itself is endowed with humor.

As to the future of the handicap theory, we cannot know whether it will take the place Dawkins has announced for it. We also cannot know, if this does take place, what the reasons for that will be. We cannot know if these reasons will be those of the theory and its explanatory and predictive power, or if it will be the contexts of justification that subtend it creating elective affinities with our mental, moral, social, and political universe.

Those who are nostalgic for a pacific vision of nature will be attentive to the morality and ontology it proposes.

Those who question the great separation will no doubt see here the promising sign of new ways of entering into relation with animals.

This will not stop the babblers from continuing to dance.

Afterword

ecently, a colleague asked me to lead a session of the seminar he organized in a prominent art school. The year's theme was representations of the animal in art. If the meetings were mostly devoted to artistic practices, he nonetheless welcomed opening up the question and addressing it to other animal specialists, and he had thought of me in recalling that I had organized the scientific committee for the exhibition *Bêtes et Hommes* at the Grand Halle de la Villette in Paris in 2007. I responded to him that I was a bit chagrined because I had excluded the term *representation* from my vocabulary since the beginning of my work. The proof of this was, I told him further, that when I was asked to prepare the exhibition that was the basis of his invitation, I had established as the *only* nonnegotiable condition that it would never bring up "our representations of the animal" nor, another way of designating this type of relation, our "beliefs" on the subject. He said that would be perfect, please come and speak to us about the reasons for which these terms pose such a problem for you.

I decided then to return to the manuscript of the book on the babbler that the memory of my successive computers had miraculously maintained throughout the last twenty-five years. I thought I would certainly find the reasons clearly spelled out there, and the seminar preparations could be finished in no time

I was astonished. The term appeared exactly twenty times, and the word *belief* stood out just as much, when I counted twenty-six occurrences of it. It is true that, in a total of 55,000 words, that is a small sum: but for a term that had *always* posed a problem, it was altogether too much.

But if I looked a little closer at it and at what interested me in the frequent use of the term with regard to the book, something else appeared: recurrent in the first twenty pages, it emerges again a few times around the middle of the manuscript and is then absent except to crop up but three times in the last quarter (one of these is in the sense not of "an image that we make" but in the political sense of democratic representation). It had thus almost disappeared between the beginning and the second part of the book.

I had therefore entertained the idea, from the moment of revisiting the manuscript to edit it, of entirely removing this term. Certainly that would have been possible. But to replace it with what? By "vision" that we have of animals? By "the idea that we fashion for ourselves about them and that prevents us from seeing them in their authentic reality"? Those were still worse! Evidently, it was not the word that was problematic; it was what it was addressed to, that strange conception of the knowledge that went along with it: the old dualism that opposes an objective exterior of facts, an authentic presence of the real, independent of us, and a continuous variability in the way that we perceive them. This is what Stengers would later teach me to recognize as what Whitehead called the bifurcation of nature.

I know now why this term was so essential. Indeed, I was already equipped (or rather encumbered) with all that had been inculcated in me during my academic training: philosophers, with just a few lovely exceptions such as the marvelous William James, do not frequently meddle with the disorder of the concrete; they are sensitive and precious, and the real will always be veiled or well covered by the hot overcoat of representations that, precisely, hinders an overly brutal exposure to things and beings themselves. But, above all, I noticed that what I was starting to really want to do, philosophy *with* animals, was at the time very badly received by philosophers.

Already, I could hear the murmurs of "that is not philosophy." Speaking in terms of "representations" could leave open the hope of being nonetheless recognized as a "philosopher," who respected the codes, who remained in her place, on the right side of the bifurcation. It amounted to proving my credentials. Except that, in any event, the demonstration of credentials would not be enough.

I was able to see in rereading it, in fact, that in the course of the book I was becoming progressively disinterested in this will to conformism and was breaking with the idea of representation. Its progressive disappearance in the course of the pages is the sign of this change.

Or, more exactly, something else captured me. For the more that the babblers are present in the text, the more their presence rekindles the writing, the more I am obligated to consider how the researchers engage in rendering them intelligible and interesting, the more I am touched by their collective intelligence (I mean to say humans and animals conjointly), the less the term has reason to be mobilized. As Isabelle Stengers posed it to me beautifully one day, "the babblers happened to me." And this type of thing that happens to you, it is an event. Seeing the babblers dance, seeing the scientists bustling about telling more stories about them that in turn enmesh them in further enigmas, these are events. They are what the book recounts.

I therefore decided, when I was revising the manuscript, to leave the "representations" as often as they appeared, to maintain the irritated (or a bit ashamed) spirit that led me to reconsider these passages, or even to remove them. And to trust in the obviousness of the progressive effacement of the term as one advances in the book, to have confidence that this obviousness would make this story rustle under the surface of the text.

We wrote to each other from time to time, Amotz Zahavi and me. He gave me news of the babblers, told me how the season had gone for them, whether the rains had been sparse and life was difficult, or rather if it had been a good season and many births had taken place, and he spoke to me of the new groups that were observed and studied. Many new things were discovered regarding the babblers. Thus we learned that they frequently use objects to communicate—we had long thought that only primates do this type of thing. When a babbler, male or female, wishes to couple with another, it will discreetly signal so by picking up a small stick, a leaf, or another such thing in its immediate environment, situate itself in the field of vision of he or she who it wishes to call out, shake the object, still discreetly, and draw the other to a remote area. The gesture is sufficiently anodyne not to be perceived by those who are not particularly watching for it. We learned many other things. One human life would not suffice to know the babblers. All the more so since they will likely not cease inventing new ways of living together, the best that they can.

Zahavi died on May 12, 2017. In August of that year he would have been eighty-nine years old. He had hoped to still take part in a last conference at the university scheduled for May 24. He was delighted that he was able to take part in teams that made new discoveries in the area. Would he have started by saying, as was his habit, "I am going to tell you things that no one believes"?

I realized that I knew very little about him. We spoke only of the birds, of the puzzles they posed, of what they teach us about the life of social beings, and this was as it should have been.

Liège, January 10, 2021

Notes

INTRODUCTION

1. [The Arabian babbler has recently been reclassified from the genus *Turdoides* to *Argya*, and from the family *Timallidae* to *Leiothricidae* based on molecular phylogenetic studies. Despret's quotation is from a 1990 article by Amotz Zahavi. The taxonomic reclassification of the Arabian Babbler occurred in 2018, after Zahavi had died in 2017. See A. Cibois, M. Gelang, P. Alström, E. Pasquet, J. Fjeldså, P. G. P. Ericson, and U. Olsson, "Comprehensive Phylogeny of the Laughingthrushes and Allies (Aves, Leiothrichidae) and a Proposal for a Revised Taxonomy," *Zoologica Scripta* 47, no. 4 (2018): 428–40; and Frank Gill and David Donsker, eds., "Laughingthrushes and Allies," *World Bird List Version 9.1* (International Ornithologists' Union, 2019). I have added the new classifications in brackets here, so that readers can see both the original classification that Zahavi was working under and the updated taxonomy. In addition, I have included a few additional sections from the Zahavi article Despret quotes, also in brackets, that bear on specific points and analyses that Despret makes in the book.—Trans.]

2. Amotz Zahavi, "Arabian Babblers: The Quest for Social Status in a Cooperative Breeder," in *Cooperative Breeding in Birds,* ed. Peter Stacey and Walter Koenig (Cambridge, UK: Cambridge University Press, 1990), 105.

3. Vinciane Despret, "Éthique et éthologie: Une histoire naturelle de l'altruisme," *Cahiers d'éthologie* 11, no. 2 (1991): 141–266.

4. Michel Foucault, *Histoire de la folie à l'âge classique* (Paris: Gallimard, 1972). [As *The Dance of the Arabian Babbler* book was originally written in French, Despret cites the French editions of French-language scholars. We have generally given the citation to the English translations of those works where they exist and we have

been able to access them, as in note 6 below with Isabelle Stengers. We maintained Despret's citation of the French edition where there is not an easily accessible English translation or where the English version is not equivalent to the French source, as in the present citation of Foucault. Despret usually cites anglophone scholarship in the English editions, but sometimes she cites the French ones. Again, we have usually given the citation to the English edition, except in those cases where she specifically analyzes aspects of the translation, such as the French translator of Darwin rendering "selection" as "election" [élection] with obvious political connotations.—Trans.]

5. I thank Lucien Francois for this concise definition of what I wanted to express.

6. We will have occasion to return to the subject of our very emotional attachment to a moral representation of the world and causality. I make a simple remark here in thinking that our epistemology is often a moral staging of our relation to the world when we think of the formulation "things should have happened this way," an expression either of determinism where things are found in a "must be" or an obligation to become, or else an aesthetic and moral vision of the obligations of the past in relation to the present (when Isabelle Stengers, in *The Invention of Modern Science* [trans. Daniel W. Smith; Minneapolis: University of Minnesota, 2000], analyzes "good causes," it seems to me that we are speaking of the same things).

7. I do not make a distinction—because it would not seem useful to me here for reasons that are too long to explain—between the context of discovery and the context of justification. I differentiate myself here from the terminology of Paul Feyerabend (*Against Method*; London: New Left Books, 1975), who uses the term justification instead to refer to the stage of research during which the researcher will "justify" and advocate for their discovery with the goal of interesting or convincing others. My reasons for not separating the context of discovery from the context of justification will perhaps become clearer over the course of the following chapters. Whatever it may be, the context of discovery, like the objects that underlie scientific productions, makes up an integral part of the context of justification, and is but one of its particular moments.

8. Robert Rosenthal, *Experimenter Effects in Behavioral Research* (New York: Appleton, 1966).

9. Stengers *Invention of Modern Science*.

10. Robert K. Merton, "The Self-Fulfilling Prophecy," *Antioch Review* 8, no. 2 (1948): 193–210.

11. [Despret's use of the term "amateur" builds on other instances where she has analyzed and mobilized the concept, and on Bruno Latour's exploration of it. Notably, it does not mean so much a naïve or uninformed practitioner, but someone who is deeply engaged, who "loves" the topic (as in the French root of the term). See Vinciane Despret, *What Would Animals Say If We Asked Them the Right Questions?*, trans. Brett Buchanan (Minneapolis: University Of Minnesota Press, 2016); and Bruno Latour, *Chroniques d'un amateur de sciences* (Paris: Presses de l'école des mines, 2006).—Trans.]

12. Rosenthal, *Experimenter Effects*, 109.

13. "I will call 'humor'," says Isabelle Stengers, "the capacity to recognize oneself as a product of the history whose construction one is trying to follow—and this in a sense in which humor is first of all distinguished from irony. . . . Humor, itself, is an art of immanence" (*Invention of Modern Science*, 65).

14. A term that I borrow from Bruno Latour (*We Have Never Been Modern*, trans. Catherine Porter; Cambridge, Mass.: Harvard University Press, 1991) and by which I designate the mediators between nature and culture.

15. Rubin's figure is a black and /or white shape that presents the characteristic of offering two images through the play of ground and figure: one can see it either as a vase (a black figure surrounded by white), or as two profiles that are facing one another (two white figures on a central black ground). It seems impossible to perceive the two images at the same time, one always appearing as the ground, the other as the figure. Once one of the two images are perceived, the other appears less easily.

1. THE THEORETICAL CONTEXT

1. Erwin Staub, "A Child in Distress: The Influence of Nurturance and Modeling on Children's Attempts to Help," *Developmental Psychology* 5, no. 1 (1971): 124–32.

2. V. C. Wynne-Edwards, *Evolution through Group Selection* (Oxford, UK: Blackwell Scientific Publication, 1986), 6.

3. It was to oppose this utilitarian–creationist conception that Darwin was sometimes preoccupied more with the failures of nature than its adaptive successes: "It may not be a logical deduction, but

to my imagination it is far more satisfactory to look at such instincts as the young cuckoo ejecting its foster-brothers, ants making slaves, the larvae of the ichneumonidae feeding within the live bodies of caterpillars, not as specially endowed or created instincts." Charles Darwin, "Instinct," in *The Origin of the Species by Means of Natural Selection of the Preservation of Favored Races in the Struggle for Life* (London: John Murray, 1859), 263.

4. According to Deleuze and Guattari, the reorganization of the function of aggression—once it turns toward individuals of the same species—"rather than explaining the territory, presupposes it." The "territorializing factor" must be seen as an emergence, "a becoming-expressive of rhythm or melody, in other words, in the emergence or proper qualities [. . .] Property is fundamentally artistic because art is fundamentally poster, placard." *A Thousand Plateaus: Capitalism and Schizophrenia*, trans. Brian Massumi (Minneapolis: University of Minnesota Press, 1987), 316.

5. Stephen Jay Gould, *Bully for Brontosaurus: Reflections on Natural History* (New York, W.W. Norton, 1991).

6. Antonello La Vergata, "Le bases biologiques de la solidarité," in *Darwinisme et société*, ed. Patrick Tort (Paris: Presses universitaires de France, 1992), 55–89.

7. A zoologist by training, Kropotkin contributed to spreading Darwinian theories in Russia at the beginning of the twentieth century. We find an exposé of his theories in a book published in French in 1906, *L'entr'aide, un facteur de l'évolution*, which was rereleased in 1979 (Paris: Editions de l'entr'aide). This book was published in English in 1902 (*Mutual Aid: A Factor in Evolution*) and had been previously published as essays in an English periodical called *The Nineteenth Century*.

8. Particularly in the television programs and reporting dedicated to the animal world.

9. Further into the book we will consider—with Paul Feyerabend—how some scientists' strategies are forms of propaganda. Also relevant here is the role of the "becoming" of the theories, since when they are widely diffused the message is reinterpreted and its content changes imperceptibly as an accent is placed on one sole aspect of a given theory.

10. Helena Cronin, *The Ant and the Peacock: Altruism and Sexual Selection from Darwin to Today* (Cambridge, UK: Cambridge University Press, 1991), 278.

11. And here we pass from one hypothesis to a more fundamental hypothesis.

12. We find this example in the two versions of the book by Richard Dawkins (*Le nouvel esprit biologique,* trans. Julie Pavesi and Nadine Chaptal [Verviers: Marabout, 1980] and the corrected version of *La gène égoiste,* trans. Laura Ovion [Paris: Armand Colin, 1989]). In English these correspond to the first and second editions of the book *The Selfish Gene* (1976; repr., Oxford, UK: Oxford University Press, 1989).

13. Amotz Zahavi, "Why Shouting?," *American Nature,* 113 (1979): 157–59.

14. William Hamilton, "The Genetical Evolution of Altruism," *Journal of Theoretical Biology* 7, no. 1 (1964): 1–52. Also see on this subject John Maynard Smith, *Problems in Biology* (Oxford University Press, 1986); and also Pierre Jaisson, *La fourmi et le sociobiologiste* (Paris: Odile Jacob, 1994).

15. S. T. Emlen, "White Fronted Bee-Eaters: Helping in a Colonially Nesting Species," in *Cooperative Breeding in Birds,* ed. Peter B. Stacey and Walter D. Koenig (Cambridge, UK: Cambridge University Press, 1990), 305–39.

16. See respectively under this heading: Emlen, "White Fronted Bee-Eaters"; H. Richner, "Helpers at the Nest in Carrion Crows Corvus corone corone," *Ibis* 132 (1990): 105–8; and H. Reyer, "Pied Kingfisher: Ecological Causes and Reproductive Consequences of Cooperative Breeding," in *Cooperative Breeding in Birds,* ed. Stacey and Koenig, 527–58; R. Curry and P. Grant, "Galapagos Mockingbirds: Territorial Cooperative Breeding in a Climatically Variable Environment," in *Cooperative Breeding in Birds,* ed. Stacey and Koenig, 289–332; S. Strahl and A. Smith, "Hoatzins: Cooperative Breeding in a Folivorous Neotropical Bird," in *Cooperative Breeding in Birds,* ed. Stacey and Koenig, 131–56.

17. Ronald Mumme, Walter Koenig, and Francis Ratnieks, "Helping Behavior, Reproductive Value, and the Future Component of Indirect Fitness," *Animal Behavior* 38, no. 3 (1989): 331–43.

18. Glen Woolfenden and John Fitzpatrick, "Florida Scrub Jays: A Synopsis after 18 Years of Study," in *Cooperative Breeding in Birds,* ed. Stacey and Koenig, 239–66.

19. It is necessary to do justice to the theory and nuance this critique: if it is true that it is in occupying oneself with one's own offspring that an organism has the greatest chances of transmitting

its genes, it is equally plausible that the helpers at the nest are for the most part unable to reproduce, either for reasons of hierarchy or saturation of habitat.

20. Emlen, "White Fronted Bee-Eaters."

21. Bruce Waldman, "The Ecology of Kin Recognition," *Annual Review of Ecology and Systematics* 19 (1988): 543–71.

22. See for example J. Philippe Rushton, "Genetic Similarity, Human Altruism and Group Selection," *Behavioral and Brain Sciences* 12 (1989): 503–59.

23. Smith, *Problems in Biology*.

24. Stephen Jay Gould and Richard Lewontin, "The Spandrels of San Marco and the Panglossian Paradigm: A Critique of the Adaptationist Programme," *Proceedings of the Royal Society of London B: Biological Sciences* 205, no. 116 (1979): 205–98.

25. This reframing renders the gene very similar to what Latour calls a black box: an establishing dispositive between the data that enter it and the data that issue from it—a relation that no scientist (at least among those who use this type of dispositive) would dream of contesting the definition of.

26. Stephen Jay Gould, *Hen's Teeth and Horse's Toes* (New York: W. W. Norton, 1983); Stephen Jay Gould, *The Panda's Thumb* (New York: W. W. Norton, 1980).

27. The concept of exaptation characterizes the attempt to surpass utilitarian reasoning of the following type: if Y is useful for X, then X is the cause of Y—in other words, the reasoning makes a quick passage from utility to cause. The concept corresponds to that of preadaptation in the sense that it indicates the fact that an organ, a structure, or a behavior is developed according to an old utility, sometimes radically different than the utility that this structure or behavior can have today. It can even be that it developed not because of its own utility but because it constitutes a lateral character (side effect) of a useful structure.

28. As Patrick Tort calls them in pointing out the singular articulation between scientific and ideological discourse; see "L'effet réversif de l'évolution," in *Darwinisme et société*, ed. Tort, 13–47.

29. A vision that leaves to the genes a large share of responsibility in behavior, and that reduces the liberties to a very limited space (like the sociobiologists who gave the field of knowledge a media-friendly and caricatural image did) seems to me to link up with an ontology

according to which essence precedes existence. Clearly, attributing to the genes the determination of certain characteristics encourages "bad faith" since in the place of assuming the liberty of becoming—I am what I do, the order of the causes is reversed—I act cowardly because I am a coward: I do what I am. It is in this sense that genetic determinism pushed to its extreme is at once both guilt-absolving and disempowering: I am not alcoholic because I drink, my genes make me an alcoholic, thus I drink.

30. I am thinking here of the sentiment of democratic melancholy which was spoken of at the time of the fall of the Berlin Wall, but which seems to me to have appeared since the 1970s. I am neither a historian nor a political scientist and my judgment is not free from the defects of caricature. But I remember those years and the void left by the stalemate of 1968. We had wanted to do what our elders had done, but had only windmills against which to tilt. Their revolution complicated the objects of our own revolution. Sociobiology, the provocative songs of Sardou, plans for the extension of military service, and abortion thus had for us a very powerful mobilizing force.

31. Cited by Michel Veuille, *La sociobiologie* (Paris: PUF, 1986).

32. Michel Maffesoli, "Une forme d'aggregation tribale," in *Faits divers: Annales des passions excessives* (Paris: Autrement, 1993), 114–27.

33. Robert Trivers, "The Evolution of Reciprocal Altruism," *Quarterly Review of Biology* 46 (1971): 35–39, 45–47.

34. Curry and Grant, "Galapagos Mockingbirds."

35. K. Rabenold, "Campylorhynchus Wrens: The Ecology of Delayed Dispersal and Cooperation in the Venezuelan Savanna," in *Cooperative Breeding in Birds*, ed. Stacey and Koenig, 157–96.

36. For the white-fronted bee-eater, see S. T. Emlen, "Cooperative Breeding in Birds and Mammals," in *Behavioral Ecology: An Evolutionary Approach*, ed. J. R. Krebs and N. B. Davies (Oxford, UK: Blackwell Publishing, 1984), 305–39. For the jay, see J. Brown and E. Brown, "Mexican Jays: Uncooperative Breeding," in *Cooperative Breeding in Birds*, ed. Stacey and Koenig, 267–88.

37. J. Ligon and S. Ligon, "Green Woodhoopoes: Life History Traits and Sociality," in *Cooperative Breeding in Birds*, ed. Stacey and Koenig, 31–66.

38. G. Wilkinson, "Reciprocal Food Sharing in the Vampire Bat," *Nature* 308 (1984): 181–84. See also G. Wilkinson, "Reciprocal

Altruism in Bats and Other Mammals," *Ethology and Sociobiology* 9 (1988): 85–100.

39. Konrad Lorenz, *On Aggression* (New York: Harcourt, Brace and World, 1966).

40. Stephen Jay Gould, *Ever Since Darwin* (New York: W. W. Norton, 1977).

41. Charles Darwin, *The Descent of Man and Selection in Relation to Sex* (London: J. Murray, 1871).

42. The example of the eye is striking in this regard: thanks to the discoveries of Helmholtz, he was able to claim, in 1871, that if an optician had sold him an instrument made with so little care, he would have judged it altogether fair to report him.

43. One could think that certain elements were able to favor the resumption in interest for choice behavior in females: according to Maynard-Smith (qtd. in Cronin, *Ant and the Peacock*) this would not be unfamiliar to the virulence of American feminist movements. I note in passing an amusing paradox—which ceases to be so (paradoxical and amusing) if we analyze it fully—that this research on female choice leads, in its logical conclusion, *to removing from these choices the idea of choice itself.* As uncomfortable as it may be to think about, we see that it is at the notion of a "choice" totally determined by necessity that theories want to drive. What they are searching for is the general law that presides over this necessity. Now, a choice determined by necessity—and I insist here "determined by necessity" refers to the strictest sense of the term determined—really has no choice, on the contrary, and even doubly, since each of these terms ("determined" and "necessity") is in total contradiction with the idea of freedom of choice. The will of the female is nothing other than a pure product of evolution, determined as it is to engender other products of evolution.

In another perspective belonging to both the sociology of the sciences and anthropology (which is not incompatible with the hypotheses mentioned above), we could think about what was happening in the "lie" and deception among animals. Research rather brutally changed its object: in the place of altruism, many researchers applied themselves to studying "deceit" behaviors among animals. Beyond the phenomena of saturation of the domain or strategies of research, we can think with J. S. Kennedy (*The Next Anthropomorphism*; Cambridge, UK: Cambridge University, 1992) that the problems of lying offered a privileged terrain for cognitivists and the theorists of consciousness and intentionality in animals. Would not the problems of choice have opened perspectives akin to a more "psychologizing" vision of ethology?

44. Roland Barthes, *Mythologies,* trans. Annette Lavers (New York: Farrar, Straus and Giroux, 1972).

45. Melvin J. Lerner, cited in J. P. Deconchy, "Systemes de croyance et représentations idéologiques," in *Psychologie sociale,* ed. S. Moscovici (Paris: PUF, 1984), 331–55.

46. Gould and Lewontin, "Spandrels of San Marco"; see also Stephen Jay Gould, *Beautiful Life: The Burgess Shale and the Nature of History* (New York: W. W. Norton, 1989); and Stephen Jay Gould, *Bully for Brontosaurus: Reflections on Natural History* (New York: W. W. Norton, 1991), on the extinction of the dinosaurs.

47. Gould, *Panda's Thumb,* 84–85.

48. Lorenz, *On Aggression,* 38 (regarding the Argus pheasant).

49. Isabelle Stengers, *The Invention of Modern Science,* trans. Daniel W. Smith (Minneapolis: University of Minnesota Press, 2000), 44–45.

50. Michel Tournier, *Le Vent Paraclet* (Paris: Gallimard, 1977).

51. Kropotkin, *L'entr'aide.*

52. See the analyses of Michele Acanforra (89–131), Antonello La Vergata (55–89), and Michael Löwy (161–68), all in *Darwinisme et société,* ed. Tort.

53. Kropotkin, *Mutual Aid,* 49.

54. A change began around the beginning of the 1980s with the entrance into laboratories and field research sites of sociologists and anthropologists of science; see Bruno Latour and Steve Woolgar, *Laboratory Life: The Construction of Scientific Facts* (Los Angeles: Sage, 1979).

55. Cited by Gould in *Bully for Brontosaurus.*

56. If this were the case, we would situate ourselves within a particular hypothesis: belief in the unity of nature. Of course, this is a view that is spread uniformly enough: it postulates that natural things are identical. The sociological analysis of Kropotkin's zoology, for example, rests on this belief since it is founded on the postulate that "all other things being equal" by which particular terrains are given as a neutral variable (since they are identical).

57. Isabelle Stengers, *The Invention of Modern Science,* trans. Daniel W. Smith (Minneapolis: University of Minnesota, 2000).

58. This definition is that of Claude Lévi-Strauss, *Anthropologie structurale* (Paris: Plon, 1974), 354, who himself uses it in his analysis of the model of Von Neumann.

59. Robert Trivers, "Parent–Offspring Conflict," *American Zoologist* 14 (1974): 249–64.

60. Some, however, escape the rule, since they do not refer to an organism but to a particular behavior: this would be the case with the strategies of the dove and the hawk, of the bourgeois, of the cheater and the pigeon, et cetera.

61. Amotz Zahavi, personal communication.

62. According to Zahavi (personal communication), the "arms race" theory is not founded on adaptation but on amelioration. A comparative reading of the literature devoted to the relations of the cuckoo with its host shows the choice and the consequences of each of the arguments presented by one or the other school.

2. RITUALS BETWEEN ALTRUISM AND REPRODUCTIVE FUNCTION

1. Julian Huxley, "The Courtship Habit of the Great Grebe with an Addition of the Theory of Sexual Selection," *Proceedings of the Zoological Society of London* 35 (1914): 491–562.

2. Desmond Morris, "Typical Intensity and Its Relationship to the Problem of Ritualization," *Behavior* 11 (1957): 1–12.

3. Barbara B. Smuts and John M. Watanabe, "Social Relationships and Ritualized Greetings in Adult Male Baboons (*Papio cynocephalus anubis*)," *International Journal of Primatology* 11, no. 2 (1990): 170, citing Roy A. Rappaport, "The Obvious Aspects of Ritual," in *Ecology, Meaning, and Religion,* ed. Roy A. Rappaport (Berkeley, Calif.: North Atlantic Books, 1979), 173–221.

4. [Here Despret has carried out a comparison of the terms in French, Latin, English, and Australian Aboriginal languages. One of her points is that the same etymological link would not seem to be present in Aboriginal languages, and that, by association, the supposed link between the material, biological organs of the body and the juridical act of swearing in the European languages is likely a false one.—Trans.]

5. Irenaüs Eibl-Eibesfeldt, *Ethology: The Biology of Behavior,* trans. Erich Klinghammer (New York: Holt, Rinehart and Winston, 1970).

6. Konrad Lorenz, *On Aggression* (New York: Harcourt, Brace and World, 1966).

7. Amotz Zahavi, "The Testing of a Bond," *Animal Behavior* 25, no. 1 (1977): 246–47.

8. The western capercaillie that I was able to observe in this situation some years ago could also, when the tension on the sandy ground was at its height and he had to defend his territory from the intrusion of other males, attack a passing female. My mentor in ethology, Jean-Claude Ruwet, who supervised my observations, emphasized the consequences that this behavior could have: the less-aggressive males with territories on the periphery could augment their chances of reproducing (with the females chased from the center), which would have as an effect the regulation of the potential aggression of the population. This hypothesis does not contradict, nor is it contradicted by, Zahavi's. It situates its explanatory level at that of final causes. It seemed to me interesting to connect the two in order to show (1) the possibility of explanations at different levels, but especially (2) the particular coloration that can affect the choice of one or the other explanatory factor.

9. The two first examples are respectively from Lorenz, *On Aggression*; and Irenäus Eibl-Eibesfeldt, *Ethology: The Biology of Behavior* (New York Holt, Rinehart, Winston, 1970). The third is Zahavi's (personal communication). Beyond the functionalism of the explanation, this interpretation of tears leads us to consider a possible meaning to the act of crying. We could consider tears as a gesture offering blindness: it is no longer the gaze that matters, but its absence. This implies a changing of perspective: the face becomes what the gaze of the other rests on. It becomes passivity, offering. The crying gaze effaces the subject to present it instead as an object of the gaze.

10. Richard Dawkins, *La gène égoiste,* trans. Laura Ovion (Paris: Armand Colin, 1989), 70.

3. THE ARABIAN BABBLER

1. This is no doubt due, rather than to the translation of behaviors in terms of information, to the heritage of the utilitarian tradition that marks Darwinian thought. From a strictly epistemological point of view, one could be shocked by this equating of what something means and what it is good for, which seems to efface the irreducibility between the causes (final in this instance) and the intention.

2. In the same way, dancing after a bath is a strange choice since it slows the time it takes to dry off.

3. Amotz Zahavi, "Arabian Babblers: The Quest for Social Status in a Cooperative Breeder," in *Cooperative Breeding in Birds,* ed. Peter Stacey and Walter Koenig (Cambridge, UK: Cambridge University Press, 1990), 129.

4. William James relates a discussion taking place around a squirrel clinging to the trunk of a tree. The entire question—which his friends were debating—was to know if the human who was trying to perceive it had or had not turned around the squirrel in knowing that "this human witness tries to get sight of the squirrel by moving rapidly round the tree, but no matter how fast he goes, the squirrel moves as fast in the opposite direction, and always keeps the tree between himself and the man, so that never a glimpse of him is caught. The resultant metaphysical problem now is this: does the man go round the squirrel or not? He goes round the tree, sure enough, and the squirrel is on the tree; but does he go round the squirrel? In the unlimited leisure of the wilderness, discussion had been worn threadbare. Everyone had taken sides, and was obstinate; and the numbers on both sides were even. Each side, when I appeared, therefore appealed to me to make it a majority. Mindful of the scholastic adage that whenever you meet a contradiction you must make a distinction, I immediately sought and found one, as follows: 'Which party is right,' I said, 'depends on what you practically mean by "going round" the squirrel.'" [From James's lecture "What Pragmatism Means," part of a series of eight lectures on "What Is Pragmatism" delivered in honor of John Stuart Mill, printed in Pragmatism: A New Name for Some Old Ways of Thinking (London: Longmans, Green, and Co, 1907), 43–81.—Trans.]

5. And it is not a coincidence, I believe, that he would refer to "writing" and the "publishable" when I spoke to him of the fact that I, too, was involved in a publication project. I think that Roni continued to want to maintain what I call here our "community of language."

6. Roni Osztreiher, "Influence of the Observer on the Frequency of the 'Morning-Dance' in the Arabian Babbler," *Ethology* 100 (1995): 320–30.

7. Amotz Zahavi, "Reliability in Communication Systems and the Evolution of Altruism," in *Evolutionary Ecology,* ed. Bernard

Stonehouse and Christopher M Perrins (London: Macmillan, 1977), 253–59.

8. Amotz Zahavi, "Communal Nesting by the Arabian Babbler. A Case of Individual Selection," *Ibis* 116 (1974): 84–87.

9. Amotz Zahavi, "Cooperative Nesting in Eurasian Birds," in *Proceedings of the 16th International Ornithological Congress, Canberra, Australia, 12–17 August, 1974,* ed. H. J. Frith and J. H. Calaby, (Canberra: Australian Academy of Science, 1976), 685–93.

10. Amotz Zahavi, "The Cost of Honesty," *Journal of Theoretical Biology* 67, no. 3 (1977): 603–5; and Zahavi, "Communal Nesting."

11. This also recalls an example from Konrad Lorenz, *On Aggression* (New York: Harcourt, Brace and World, 1966), relating the conflict that can happen between two spouses among pigeons, songbirds, and parakeets, when the female wants to feed the male. Feeding the other, Lorenz suggests, is not only a social need, but at the same time a privilege reserved to the individual of higher rank.

12. This is Jonathan Wright. To be entirely precise, we must mention the fact that Jon now works at Cambridge, after many years at Oxford. We nevertheless call him, following Zahavi, the Oxford zoologist, because he presents, according to us, many Oxfordian characteristics—at least in his theories and practices.

13. This distinction has not always been understood and has been the source of confusion. Zahavi clarifies it using the example of the professor: the relation that ties her to students is clearly hierarchical and is determined early in the relationship. However, the status of the professor depends on numerous factors tied to her own qualities. And on this status will depend how the professor is treated and considered: a good status will correspond to respect and deference from the students. On the contrary, a professor with a bad status will not be listened to nor respected, but will instead be met by chaos, laughs, and outbursts. It should be noted that in his later writings Zahavi opted for the term "prestige."

14. I use the concept in reference to J. L. Austin's theory of speech acts, and more specifically in reference to performative acts, to take account of both the exhibition (display) element and the dynamic aspect of this parade, because it can, in its very manifestation, change the status of the one who performs it. We could see a parallel here with the paradigm that Shirley Strum and Bruno Latour speak of

in their article regarding social ties among baboons, "From Baboon to Human," *Social Science Information* 26, no. 4 (1987): 783–802. In fact, it amounts to a happy coincidence. I did not know of their concept when I wrote the first version of this manuscript. The paradigm set out by Strum and Latour describes the conceptual frame that allows the definition of society as the product of creativity of the social actors themselves. Thus baboons actively construct their societies through arranging it in the course of complex interactions. This definition does not match that of performative exhibition. However, the analysis of a babbler society shows that performative exhibition, as a tool of compromise, is the mode or arrangement for a society created by the actors themselves. The two concepts, then, are not strangers.

15. Bernd Heinrich, *Ravens in Winter* (London: Barrie and Jenkins, 1990).

16. Trygve Slagsvold, "The Mobbing Behavior of the Hooded Crow *Corvus corone cornis*: Antipredation defence or self advertisement?," *Fauna norv. Ser. C, Cinclus. Supplement* 7: (1984): 127–31.

17. One imagines with laughter the cartoon scenes that this would give rise to if it were not the case: a predator yelling out "And me, what am I doing here?" standing in front of ravens in battle. I relate this because the laughter, in its Bergsonian function, puts a finger on precisely one of the flaws of this type of experimentation: it consists of mechanisms plastered over the living.

18. The term used to designate the fact that many birds join together and can even harass or attack an animal that threatens them.

19. Lorenz, *On Aggression,* 36–37; Amotz Zahavi, "Mate Selection: A Selection for Handicap," *Journal of Theoretical Biology* 53 (1975): 209.

20. Richard Dawkins, in *The Selfish Gene,* trans. Laura Ovion (Paris: Armand Colin, 1976), says that she must not confuse big muscles and false shoulder pads put under a jacket.

21. Zahavi, "Mate Selection."

22. "The expressive is primary in relation to the possessive; expressive qualities, or matters of expression, are necessarily appropriative and constitute a having more profound than being. . . . Messiaen is right in saying that many birds are not only virtuosos but artists, above all in their territorial songs (if a robber 'improperly wishes to occupy a spot which doesn't belong to it, the true owner sings and sings so well that the predator goes away. . . . If the robber

sings better than the true proprietor, the proprietor yields his place')." Gilles Deleuze and Felix Guattari, *A Thousand Plateaus: Capitalism and Schizophrenia,* trans. Brian Massumi (Minneapolis: University of Minnesota Press, 1987), 316–17.

23. Amotz Zahavi, "The Fallacy of Conventional Signalling," *Philosophical Transactions of the Royal Society of London B: Biological Sciences* 340, no. 1292 (1993): 227–30.

24. We note in passing that, in all the literature reviewed, no author placed these terms in quotation marks. It seems that the consensus around the technical sense is sufficiently established that it no longer poses a problem. This deserves a reflection that I cannot go into here. Nevertheless, I will keep the quotation marks in addressing an audience outside of that particular technical domain. I could be wrong in this, in the respect that many other terms in my study might also deserve to also be placed in quotation marks, and I accept in advance the well-deserved critique of "double standards."

25. See below for the prey–predator relation in which these signals also present a value.

26. Amotz Zahavi, "Decorative Patterns and the Evolution of Art," *New Scientist* 80 (1978): 182–84. Symmetry would be a criterion of choice on the part of females. The anomalies that affect it would be the result of stress in the course of ontogenesis. Lipstick for women plays the role, according to Zahavi, of showing the mouth's perfection.

27. Amotz Zahavi, "The Theory of Signal Selection," in *Proceedings International Symposium on Biological Evolution,* ed. V. P. Delphino (Bari, Italy: Adriatica editrice, 1987), 307.

28. Dawkins, *La gène égoiste,* 171.

29. We have only considered here some of the examples to which this theory is thought to be able to apply its explanatory power. This explanatory power extends to all living beings, including cellular exchanges at the heart of the organism. Zahavi later worked on the application of this theory to chemical signaling, notably among messenger peptides in the reproductive cycle of yeast.

4. MODELS AND METHODS

1. Vinciane Despret, "Ecology and Ideology: The Case of Ethology," *International Problems* 63, nos. 3–4 (1994): 45–61; Vinciane Despret, "L'idée de société en éthologie," *Cahiers d'éthologie* 13, no. 4

(1995): 435–68; Vinciane Despret et al., *L'homme en société* (Paris: PUF, 1995).

2. Isabelle Stengers, *The Invention of Modern Science,* trans. Daniel W. Smith (Minneapolis: University of Minnesota Press, 2000), 75.

3. Shirley Strum and Bruno Latour, "From Baboon to Human," *Social Science Information* 26, no. 4 (1987): 783–802.

4. Strum and Latour, 787.

5. Strum and Latour, 789.

6. Strum and Latour, 790.

7. Konrad Lorenz, *L'envers du miroir* (Paris: Flammarion, 1975), 16.

8. I thank Ezio Tirelli for having emphasized this distinction in a discussion on the subject.

9. [It should be noted here that French shares one term, *expérience,* for "experiment" and "experience," and that this sentence plays off of that word's internal resonance. While the lack of a hard and fast distinction may seem strange to readers of English, it also bears noting that the crossover in French serves to de-technicalize the concept of the experiment and highlight that it is, indeed, a form of focused or controlled experience.—Trans.]

10. Amotz Zahavi, "Some Comments on Sociobiology," *Auk* 98 (1981): 414.

11. Stengers, *Invention of Modern Science*.

12. A number of controversies in animal psychology will elsewhere bear on this argument (Harry Collins and Trevor Pinch, *The Golem: What Everyone Should Know about Science*; Cambridge, UK: Cambridge University Press, 1994). This procedure seems rather old since we already find it at the beginning of the last century in Kropotkin. Indeed it rises up against the assertion according to which marmots are aggressive: it is only in the conditions of captivity that the phenomenon can take place, the marmots that he met in the field were, for their part, quite peaceful.

13. We can also find a detailed tracking of this in the book by Marian Dawkins (*Through Our Eyes Only*; Oxford, UK: Freeman Spektrum, 1993) that uses not the procedure, but rather the data coming out of the experiment to discuss the problem of the attribution of cognitive faculties among the birds.

14. In an article before 1973 dedicated to the hybridization of some birds, and in some articles written in collaboration with the

a priorists on the behavior of the cuckoo, for example, the results of research carried out with a Japanese zoologist: A. Lotem, N. Hitochi, and A. Zahavi, " Rejection of Cuckoo Eggs in Relation to Past Age: A Possible Evolutionary Equilibrium," *Behavioral Ecology* 5 (1992): 128–32.

15. Robert Slotow, Joe Alcock, and Steven Rothstein, "Social Status Signalling in White Crowned Sparrows: An Experimental Test of the Social Control Hypothesis," *Animal Behaviour* 46 (1993): 977–89.

16. Amotz Zahavi, personal communication.

17. J. S. Kennedy, *The New Anthropomorphism* (Cambridge, UK: Cambridge University Press, 1992), citing Michael W. Fox, *The Whistling Hunters: Field Studies of the Asiatic Wild Dog* (Albany: SUNY Press, 1984), who in turn is citing Alf Wannenburgh, *The Bushmen* (New York: Mayflower Books, 1979).

18. "The authority of experimental science, its pretension to objectivity, has only a negative source: a statement–from a given time, of course, and not in the absolute–the means of demonstrating that it is not a simple fiction, relative to the intentions and convictions of its author" (Stengers, *Invention of Modern Science,* 103).

19. Zahavi, personal communication.

20. Zahavi, personal communication

21. Richard Dawkins, *La gène égoiste,* trans. Laura Ovion (Paris: Armand Colin, 1989), 313.

22. Claude Lévi-Strauss, *Anthropologie structurale* (Paris: Plon, 1974), 209.

23. [This refers to the mise en abîme, or infinite reflection, represented in paintings like Velázquez's *Las Meninas,* which famously depicts characters outside the frame of the painting through showing a mirror in the back of the scene painted that reflects toward the viewer.—Trans.]

24. Stengers, *Invention of Modern Science,* 79.

25. One will recognize here the borrowing of a title from Isabelle Stengers: *La volonté de faire science* (Paris: Les empêcheurs de penser en rond, 1990).

5. NARRATIVES AND METAPHORS

1. Stephen Jay Gould, *Hen's Teeth and Horse's Toes* (New York: W. W. Norton, 1983), 35.

2. Amotz Zahavi, "Mate Guarding in the Arabian Babbler,"

Proceedings of the International Ornithological Congress (Ottawa) (1988), 420–27.

3. Zahavi, 424.

4. Zahavi, 425.

5. Zahavi, 425.

6. Paul Feyerabend, *Against Method* (London: Verso, 1993), 114.

7. Feyerabend, 121.

8. "Immunology as a Cognitive Science," Weizmann Institute of Science, Rehovot, Israel, April 17–21, 1994.

9. Amotz Zahavi, "The Pattern of the Vocal Signal and the Information They Convey," *Behaviour* 80 (1982): 1–2. We can recall here that it is exactly the same arguments that William James uses to show the peripheral origin of the emotions that has influenced all the research on the expression of emotions.

10. Amotz Zahavi, "The Theory of Signal Selection," in *Proceedings International Symposium on Biological Evolution,* ed. V. P. Delphino (Bari, Italy: Adriatica editrice, 1987), 315.

11. Amotz Zahavi, "Reliability in Communication Systems and the Evolution of Altruism," in *Evolutionary Ecology,* ed. Bernard Stonehouse and Christopher Perrins (London: Macmillan, 1977), 258.

12. Amotz Zahavi, "Why Shouting?," *American Naturalist* 113, no. 1 (January 1979): 156; see also Thomas C. Schelling, *The Strategy of Conflict* (New York: Oxford University Press, 1963).

13. Amotz Zahavi, Ritualization and the Evolution of Movement Signals," *Behaviour* 72 (1980): 78.

14. Zahavi, "Theory of Signal Selection," 316.

15. Zahavi, "Pattern of the Vocal Signal," 5.

16. Amotz, Zahavi, "The Testing of a Bond," *Annals of Behavior* 25, no. 1 (1977): 246–47.

17. Zahavi, "Theory of Signal Selection," 320.

18. Amotz Zahavi, personal communication.

19. We could also analyze the passages as taking inspiration from the model proposed by Francis Pire in a nice analysis of Lorentzian anthropology (in an unpublished article called "Lorenz and the Pre-Programmed Human"). Drawing on Mauron's psycho-critical method, Pire proposes to analyze and discover the personal myth of the scientist by revealing the particularities of his discourse considered as a literary text: his writing tics, quirky repetitions, and surprising lacunae. The analysis is completed with more factual elements, which

are the behaviors preferred by the author in organizing his own personal myth.

20. Bruno Latour, *We Have Never Been Modern,* trans. Catherine Porter (Cambridge, Mass.: Harvard University Press, 1993), 11–12.

6. MODELS AND FICTIONS

1. Karl Marx, letter to Friedrich Engels from June 1862, in *Lettres sur les sciences de la nature (et les mathématiques),* by Karl Marx and Friedrich Engels, ed. and trans. Jean-Pierre Lefebvre (Paris: Éditions sociales, 1973) (in English, *Marx and Engels Collected Works*; London: Lawrence and Wishart, 1973, p. 380).

2. Friedrich Engels, letter to Piotr Lavrov from November 1875, in Marx and Engles, *Lettres sur les sciences* (in English, "Letter from Friedrich Engels to P. L. Lavrov, 12 November 1875," *Labour Monthly,* July 1936, pp. 437–42).

3. Steven Rose, Richard C. Lewontin, and Leon J. Kamin, *Not in Our Genes: Biology, Ideology, and Human Nature* (1984; repr., New York: Penguin, 1990), 242.

4. John Maynard Smith, "Game Theory and the Evolution of Behaviour," *Proceedings of the Royal Society of London: Series B, Biological Sciences* 205, no. 1161 (1979): 475–88.

5. Preface to French edition of Shirley Strum, *Almost Human* (*Presqu'humain*; Paris: Eschel, 1990).

6. R. C. Lewontin, Steven Rose, and Leon J. Kamin, *Not in Our Genes: Biology, Ideology, and Human Nature* (New York: Pantheon, 1984).

7. Amotz Zahavi, "The Theory of Signal Selection," in *Proceedings International Symposium on Biological Evolution,* ed. V. P. Delphino (Bari, Italy: Adriatica editrice, 1987), 310.

8. Amotz Zahavi, "The Fallacy of Conventional Signalling," *Philosophical Transactions of the Royal Society of London B: Biological Sciences* 340, no. 1292 (1993): 227.

9. Bruno Latour, *Nous n'avons jamais été modernes: Essai d'anthropologie symétrique* (Paris: Éditions La Découverte, 1991), 254.

10. [In an interesting animalized difference, the French term for "crowbar" (*pied-de-biche*) would translate into English as "deer's foot." There also seems to be no ready alternative in English.—Trans.]

11. Amotz Zahavi, personal communication.

12. Jonathan Wright, personal communication.

13. Mark Kirkpatrick, "The Handicap Mechanism of Sexual Selection Does Not Work," *American Naturalist* 127 (1986): 222–40.

14. Shirley Strum expressed her discouragement and her anger in the face of what she called the "seizure of power" of the models over reality. When she came back from the field where she observed *Papio anubis* baboons, she did so with data that seriously challenged the theories of dominance among primates, and specifically among baboons. She found herself confronted with a veritable circling of the wagons. One of the arguments her colleagues made was to tell her that the mathematical models of game theory contradicted, in a clear manner, the strategies that her observations seemed to reveal to her. See *Almost Human: A Journey into the World of Baboons* (New York: W. W. Norton, 1987).

15. Heinz Richner, "Assessment of Expected Performance and Zahavi's Notion of Signal," *Animal Behaviour* 45 (1993): 399–401. For the response, see Amotz Zahavi, "Some Clarifications of the Notion of Signals: A Reply to Richner," *Animal Behaviour* 45 (1993): 402.

16. Amotz Zahavi, "Some Comments on Sociobiology," *Auk* 98 (1981): 414.

17. Richard Dawkins, *The Selfish Gene* (Oxford, UK: Oxford University Press, 1976), 159–60.

18. Richard Dawkins, *The Selfish Gene* (1976; repr., Oxford, UK: Oxford University Press, 1989), 308.

19. Alan Grafen, "Biological Signals as Handicaps," *Journal of Theoretical Biology* 144 (1990): 517.

20. A term that itself is also the object of a borrowing here since it is the title of a book edited by Isabelle Stengers (*D'une science à l'autre: Les concepts nomades*; Paris: Éditions du Seuil, 1987).

21. Zahavi, "Some Comments on Sociobiology."

22. Zahavi, personal communication. I would add, in the vein of arguments that show Zahavi's total disinterest for playing the game, that when I asked him if he had been happy about Dawkins's reversal and pointed out that it was certainly a good thing that Dawkins had said that he was right, he laconically responded to me: "Why should I be happy? I know that I am right."

23. Dawkins, *Selfish Gene* (1989 ed.), 308.

Vinciane Despret is professor of philosophy at the Université de Liège and professor of ethology and sociology at the Université libre de Bruxelles. Her books include *Women Who Make a Fuss* (with Isabelle Stengers), *What Would Animals Say If We Asked the Right Questions?*, and *Our Grateful Dead: Stories of Those Left Behind*, all published by the University of Minnesota Press.

jeffrey bussolini is codirector of the Center for Feline Studies / Avenue B Multi-Studies Center, associate professor of sociology and anthropology at the City University of New York, and researcher at the Scuola di interazione uomo-animale in Bologna, Italy. She is coeditor and cotranslator of three volumes on philosophical ethology.